Cochrane Shipbuilders
Vol 3 : 1940 - 1993

by

Gilbert Mayes and Michael Thompson

Front cover : A broadside view of the tug **Powerful** *(yard no.150) being launched on 7 March 1988.*

(Roy Cressey collection)

Back cover : The **Eliza PG** *(171), the yard's penultimate newbuilding, is eased towards the fitting out berth following her launch on 26 November 1991.*

(Roy Cressey collection)

Above : The **Boston Sea Lance** *(108) meets the River Ouse just seconds after leaving the slipway on 12 July 1979.*

(Cochrane archive)

FOREWORD

With the completion of Volume 3, we have time to reflect on the considerable amount of time and effort that has gone into compiling the work, and thank all the people who have made contributions. The ground work was begun by Tony Lofthouse, David Newton and I, and then taken forward by Gilbert Mayes and Bernard McCall. This volume highlights the challenges facing the British shipbuilding industry during the post war era. These included the gradual move from steam power to diesel engines, advance in trawler, tug and coaster design and the emergence of the offshore oil industry. The Cochrane family and subsequent owners constantly were able to remain competitive through investment in new technology, ably assisted by a highly skilled drawing team and a loyal local workforce. To sum up, over the years the Cochrane yard created a reputation for the high standard of the vessels they built, and this article from a Ross Group brochure illustrates their pride in being involved with the shipyard at the time.

"Cochrane's experience and skill are second to none in their own sphere of shipbuilding. They have built over 1,500 vessels of all types, specialising in minor warships, trawlers, cargo vessels, barges, ocean going tugs, small river tugs, yachts and special craft. With an accent on quality rather than quantity, Cochrane's rarely build to standard designs but to customers' particular requirements, resulting in a succession of commissions from individual customers. Cochrane-built ships operate in many countries around the world: tough durable vessels with the quality of construction and design proved daily on all seas and all conditions of service".

Having sailed aboard the trawlers *Cape Crozier* (1334), *Lord Essendon* (1158), *Lord Ancaster* (1337) and *Ross Kelly* (1408), I can endorse these sentiments.

Michael Thompson   Hull   January 2016

ACKNOWLEDGMENTS

As noted in the Foreword, many people have contributed to the book. We wish to thank specifically all the photographers who have made their work available. We should also thank Arthur Credland for his contribution to the yard history; Bill Blow for the funnel designs, John Adamson for his personal knowledge of the yard; Iain McCall for the pre-press tasks; and Gomer Press for their work in producing the finished book. We must also pay special tribute to Ken Marshall, the yard's final managing director. He ensured that the company archives were preserved and maintained a keen interest in the three volumes of this yard history.

Published by Bernard McCall, 400 Nore Road, Portishead, Bristol, BS20 8EZ.
Website : www.coastalshipping.co.uk.   Telephone/fax : 01275 846178
E-mail : bernard@coastalshipping.co.uk.
All distribution enquiries should be addressed to the publisher.

Printed by Gomer Press, Llandysul Enterprise Park, SA44 4JL.
Telephone : 01559 362371   Fax : 01559 363758   E-mail : sales@gomer.co.uk
Website : www.gomerprinting.co.uk

ISBN : 978-1-902953-75-5

The Cochrane Shipyard

1945 - 1993

War against Germany was declared on 3 September 1939 following the German invasion of Poland two days previously and the next six years saw an intense period of activity at Selby fulfilling orders for the Admiralty and Ministry of War Transport. The directors, workforce and their suppliers responded magnificently to the challenge, often under very trying circumstances, particularly during air raids which targeted Hull. The news that the Cochrane-built Fleetwood trawler *Davara* (517) * had been shelled and sunk by a U-boat off Tory Island, Co. Donegal, on 13 September 1939 brought home the reality of war; she was the first British trawler of many to be lost to enemy action.

Requisitioned trawlers did an outstanding job as minesweepers, patrol and boom working vessels, as well as employment on the balloon barrage and many other miscellaneous naval duties. Sturdy and reliable they could withstand not only the weather but a considerable amount of punishment from enemy attack. A number of Cochrane-built trawlers used in the Norway campaign were sunk in April 1940 following air attack. These were the Grimsby trawlers, named after famous cricketers, *Larwood* (1148), *Hammond* (1149), *Jardine* (1151), sunk by own forces as a result of air attack, and *Bradman* (1167) as well as *Cape Siretoko* (1203) and all were subsequently salved by the Germans and saw service in the Kriegsmarine.

Trawlers and drifter/trawlers with their powerful winches and strong construction were particularly suited to minesweeping and the sweeping of moored mines in particular. Many were converted to use either the 'A' sweep in which two or more vessels were involved towing the serrated sweep wire between them or the Oropesa or 'O' sweep for single boat sweeping which had been developed from trials carried out in WWI by the requisitioned trawler *Oropesa* (620). The latter method allowed sweepers to clear a minefield from one side, all ships being able to work in clear water. Others were fitted out to deal with influence mines using the 'LL' electric cable sweep for magnetic mines or the bow mounted acoustic hammer box for noise detonated mines. The object of all minesweeping was to ensure a safe passage for friendly and neutral shipping by establishing clear water free from mines.

Between 26 May and 3 June 1940, the British Expeditionary Force, French and Belgian troops who had been driven back and surrounded in France and had no other means of escape were evacuated from the ports and beaches of Boulogne, Calais and Dunkirk by many small ships and boats. Over fifteen Cochrane-built ships were involved, several being damaged and one, the Marine Française trawler *Marguerite Rose*, ex *Margaret Rose* (1100), was bombed and abandoned before sinking. Also involved was the 1905-built London tug *Java* (353) which rescued some 270 troops from the Dunkirk beaches. The *Empire Henchman*, ex *Karl* (1206), made three crossings, the first towing the sailing barge *Aidie* carrying stores and the final one towing a barge with provisions and ammunition.

From the outbreak of war the shipyard was heavily involved in the construction of ships for the war effort. The first two Tree class trawlers, *Ash* (1208) and *Bay* (1209), based on the design of HMS *Basset*, had been ordered in June 1939 and the pair were to be the forerunners of a variety of minesweeping (M/S) and antisubmarine (A/S) vessels ordered from Cochrane based on trawler designs. Immediately after war had been declared the shipyard received orders for two Dance class A/S escort trawlers *Fandango* (1214) and *Foxtrot* (1215) and these were to be followed by the three Shakespearian class (1220-1222) and nine Fish class based on the *Gullfoss* (1058) for Consolidated Fisheries Ltd, Grimsby. Two further vessels, *Mackerel* (1246) and *Turbot* (1254), were completed as the controlled minelayers *Corncrake* and *Redshank* respectively.

The Isles class was by far the largest trawler derived class constructed in WWII with 145 vessels launched by various shipyards at home and abroad. Of the twenty-three Selby-built Isles class, six were completed as danlayers which carried dan buoys to mark the swept channels. An outstanding testimony to the quality of Selby and Hull workmanship was demonstrated by the Isles class trawler *Eday* (1235) which sailed 70,000 miles without refit. Before the building of these trawlers started, the shipyard received orders for a series of tugs.

Experience in WW1 had shown that the convoy system worked and many vessels damaged by the enemy could be saved if a tug was included in the escorts. The well-proven design of the prewar Cochrane-built tugs *Neptunia* (1194) and *Salvonia* (1197) for Overseas Towage & Salvage Co Ltd was developed to become the Assurance class ocean rescue tugs and twenty-one units were constructed at Selby, the class name being adopted from the first build *Assurance* (1216). Of 700 tons displacement, oil fired and providing 1350ihp they were well suited to ocean salvage and rescue and *Prosperous* (1251) travelled over 17,000 miles and undertook eleven rescues in a single year, without loss of life.

Such was the success of this class that a further class of six tugs, known as the Envoy class, was designed

* The number in brackets after the name is the yard number.

and built for the Ministry of War Transport and all were commissioned from completion in the Royal Navy. Of 868 tons displacement, oil fired and with machinery developing 1700ihp they proved, both during and after the war, to be very useful in a variety of roles.

A need was also identified for smaller harbour/coastal tugs and designs from a number of shipyards were chosen. At Selby the **Hoedic** (1105), built in 1931 for Cie Nazairienne de Remorquage et de Sauvetage, St Nazaire, was chosen as a prototype for a class of coastal tugs. Oil fired and developing 850ihp, 105.2 feet between perpendiculars and 26.6 feet beam they displaced 425 tons; eighteen tugs of this class were launched between 1942 and 1946. **Empire Fairy** (1243) being the first and **Empire Simon** (1308) the last of the class. Defensively armed and originally designed for a crew of eight men, this was often increased depending on deployment. Two further tugs in this category based on the 1936 design of the **Hornby** (1163) and known nominally as the Modified Hornby class were built; they were **Empire Darby** (1261) and **Empire Joan** (1262).

The third class of tugs built at Selby was the Stella class. Designed for long tows, where displacement and seakeeping were more important than bollard pull, they had increased freeboard and bunker space. Oil fired, engines were supplied from surplus US stock and developed about 525ihp, but they were not easy to maintain. Eight of these were built, starting with **Empire Flora** (1297) and a further two, **Empire Doreen**, launched as **Empire Hedda**, (1314) and completed as **Atlas**, and **Empire Juna** (1315) were Improved Stella class with 800ihp machinery by Amos & Smith.

Cochrane-built tugs and many trawlers were involved in Operation Neptune - the Allied D-day landings on the Normandy coast in June 1944 as well as in other Allied landings in Africa, Sicily and Italy. Individual vessels involved, too numerous to mention here, are recorded in the yard list history profiles.

With the formal surrender of the Japanese to US and British forces on 2 September 1945, World War II came to an end but the Selby shipyard still had orders to complete and had six tugs of the Stella, two Improved Stella and two of the Hoedic class still in hand besides two dry cargo motorships. These vessels, **Empire Mayring** (1305) and **Empire Mayrover** (1306), intended for Far East deployment, were of the standard 'C'-type dry cargo coasters with bridge sited well forward enabling long items of cargo to be carried either in the hold or on deck. Both were completed in 1946 and sold on by the Ministry of Transport to agents on behalf of Hong Kong based owners.

As a demonstration of the quality of both ship and crew during the hostilities, the **Kenrix** (701), built in 1921 for Robert Rix & Sons Ltd, Hull, and under the command of Capt George Simison, sailed in 300 convoys and endured 250 air attacks, shelling from long-range gun, and assaults by E-boats and U-boats. She was sold in 1945 and broken up in 1953.

In the shipyard there had been many changes and just as the founder Andrew Cochrane Snr had died in the period of the Great War so Thomas M Cochrane died during World War II on 31 May 1943. Andrew Cochrane Jnr had died in the interwar years on 14 March 1925, having been awarded an OBE on 1 July 1918 for his work and the contribution made by the shipyard to the war effort. Of far reaching effect was the introduction of 'pay as you earn' (PAYE) in 1944, a tax system which posed significant challenges to many employers and workers who had previously never come in contact with the tax system.

The business at Selby continued in the hands of the Cochrane family. The sons of Andrew Cochrane Jnr, Andrew Lewis Cochrane (who had worked at the Grangemouth Dockyard Co Ltd, Grangemouth, from 1911 to 1914), Donald McGill Cochrane and Sidney John Cochrane (after a period with Jos T Eltringham & Co Ltd at South Shields) were all directors, along with the naval architect, Victor Gray. Electric welding which had been used extensively during the war period, especially for shell plate butts and deck structures, was set to be further exploited in the post war years leading to shorter build times and other efficiencies.

After the war - a trawler boom for home and abroad

Towards the end of the war in 1945, the Ministry of Food, concerned about the nations food supply, granted licences for twelve large trawlers to be built. Basil Parkes applied to build two 180ft oil-fired trawlers for the distant water grounds and Leslie Marr to build a sister ship; the orders were placed with Cochrane. All three ships were built as coal-burners because the Ministry of Fuel was fearful that ongoing problems in the Middle East would be prolonged and may affect the supply of oil. Completed in early 1946, **St Bartholomew** (1309), and **St Mark** (1319) were the first vessels built by Cochrane on a variable price contract, one that took account of possible increases in wages, steel and machinery; **Northella** (1311) was built on the traditional fixed price contract and in the end proved to be the more costly. Much to the chagrin of the Hull owners, Owen Hellyer in particular, licences were granted, seven to Basil Parkes, three to Leslie Marr and one each to Walker Steam Trawl Fishing Co Ltd and Richard Irvine & Sons Ltd, both of Aberdeen.

Basil Parkes in his book *Trawlings of a Lifetime* - (1991), wrote: "I received a letter from Lewis Cochrane. Enclosed was a cheque for £5,000; they had just got their audited figures, he explained, and the profit they had made on the building of *St Bartholomew* had proved higher than anticipated, and he felt he should return this excess to us. A smaller cheque was sent later for the *St Mark*. What a tribute to the Cochranes' integrity and honesty; they were, perhaps, a little old-fashioned in their design work, but it is to their credit, also, that I do not think they ever built a bad ship."

With the three large trawlers already booked a further boost was received from the Icelandic Government. During the war Icelandic fishing vessels and carriers braved attack from German naval and air forces to land their catches and fish cargoes in Britain. As well as providing vital food supplies they rescued hundreds of Allied seafarers whose vessels had been sunk in the waters between Iceland and Britain.

Many of the Icelandic trawlers required replacement after the war and a committee representing British trawler builders was formed under the chairmanship of Lewis Cochrane to tender for these vessels. This resulted in a contract for thirty-two trawlers to be built in various yards, including eight at Selby (1320-7). Oil-fired and to a very high standard of fit-out the contract was completed in 1948. For his efforts on behalf of the Icelandic Government, Lewis Cochrane was awarded the Icelandic Grand Cross of the Order of the Falcon.

The dreadful winter of 1946-47 throughout Britain, with heavy snow and severe icing was followed by a rapid thaw which resulted in flooding of the yard when the River Ouse overflowed the bank defences. Ultimately this was not a major economic loss to the company and lead to the raising of the flood bank. Of more importance was the loss of many early plans, records and photographs from the foundation of the company. A photograph of the trawler *Elliði* (1325), secured alongside after launching, shows a bleak scene with deep snow. The bad weather of 1947 also coincided with the introduction of a 44 hour working week.

The wintry scene described above.

(Author's collection)

Problems encountered in the development of motor trawlers, not least in the supply of engines and winches, resulted in many owners remaining faithful to the well-proven triple expansion steam engine. Completion of the first Selby-built motor trawler, *Milford Viscount* (1319) for the Milford Steam Trawling Co Ltd, Milford, was delayed by the late delivery of the main engine and the severe weather in 1946/47 already mentioned.

With five further distant water trawlers (1328-1330 & 1334-1335) booked under the Ministry of Agriculture & Fisheries licence scheme, all five to be oil fired, the company successfully tendered for two motor drifters to be built for Lowestoft owners. Between 1903 and 1931 the Selby yard built some 70 drifters but with the rapid decline in fish stocks, drift net fishing became unviable. The pair *Dauntless Star* (1331) and *Sunlit Waters* (1332) completed in early 1948 were strengthened so that they could quickly convert to white fish trawling.

With oil fired vessels considered to be the better option it is perhaps surprising that contracts were signed in 1947 to build three coal fired trawlers for Hull owners, though the Ministry of Fuel was again involved in the decision making; more understandably one vessel for South Africa was always intended to be coal-fired. The pair for Lord Line, *Lord Wavell* (1336) and *Lord Ancaster* (1337), lasted less than two years before being converted to burn oil fuel, while the final coal burner to be built by Cochrane, *Boynton Wyke* (1343) completed in mid-1949, did not convert to oil until 1955. All subsequent steam trawlers ordered from Selby were oil-burners and many prewar trawlers were converted to burn this fuel as heavy oil supplies became more assured and available in Britain.

In October 1948, R J Coles was appointed assistant yard manager and in 1951 there were changes in the shipyard management, A C Kelly became chief draughtsman with R W Walker appointed as his assistant. Workers at the yard, in line with other shipyards, were granted two weeks paid holiday each year. Lewis Cochrane, a major figure in the British shipbuilding industry, was appointed President of the Shipbuilding Employers' Federation and in 1953 he was made an OBE for his services to the industry.

The Sea Fish Industry Act - 1951 was given Royal Assent on 10 May 1951 and within it came the setting up of the White Fish Authority (WFA). The remit of the WFA was to "reorganise, develop and regulate the white fish industry" and its principal revenue was derived from statutory levies on the first hand purchase of fish. The WFA also took over the loans & grants scheme which had been set up in 1946 to assist the owners of small fishing vessels to make improvements in efficiency, safety and more economical operations. In 1953 the loans and grants scheme was extended for the purchase of inshore, near and middle water fishing vessels and the conversion of steam vessels from coal to oil.

The scheme applied to fishing vessels less than 139 feet registered length, a scale that benefitted small shipyards such as Cochrane which established a fine reputation for the design of near and middle water trawlers. The scheme allowed trawler owners to take a grant of 25% or 30% for skipper/owners, and a loan of 85% at very low interest rates. In the development of the sea fishing industry, the WFA coordinated the science and assessment of fish stocks and the promotion of better practice in the care of fish onboard, and, on a national level, the eating of fish.

The owners in Fleetwood and Grimsby were quick to take advantage of the loans and grants scheme but as it only allowed the vessel to make two trips to the distant water grounds per year it was of little benefit to the Hull trawler owners with their emphasis on those grounds and who in fact paid the bulk of the levy. Pressure was put to bear on the government and in 1961 the WFA scheme was extended to trawlers destined to fish further afield. The first motor trawler built at Selby with the aid of a grant was the **Fleetwood Lady** (1394) for Fleetwood, followed by **Olivean** (1397) for Grimsby owners, both completed towards the end of 1954 with no delays thanks to the owners' close cooperation with the engine builders. Companies in all the major fishing ports took advantage of the grants and loans and many motor trawlers were built by Cochrane as a result.

*The **Olivean** was the first vessel built at the Cochrane yard with the aid of a grant from the White Fish Authority.*

(Cochrane archive)

During this period in 1949 and the early 1950s the Selby yard was successful in tendering for several tugs, the majority motor including the **Nethe** (1348) and **Senne** (349) for Belgian owners. An order for three steam tugs was placed by the Alexandra Towing Co Ltd, Liverpool, two to be oil-fired, and **Wallasey** (1390) which was completed as a coal burner.

Having gained considerable experience in the operation of the distant water motor trawler **Lammermuir** (729grt/1950) (H105) built by Lewis at Aberdeen, the Parkes family ordered two large motor trawlers from Cochrane, **Princess Elizabeth** (1380) and **Prince Charles** (1381), the latter being completed in January 1953. Just before Christmas 1955, **Prince Charles**, homeward from the Barents

Sea grounds and with Norwegian pilots onboard, stranded on Karken Island north of Hammerfest in a snowstorm; nine crew and a Norwegian pilot were lost. The vessel was salved and later towed back to Hull for repair having been abandoned to the underwriters. Rebuilt, she re-entered service as **Loch Melfort**. On 10 March 1957, when outward for Norwegian coast grounds in thick fog, she was in collision when 65 miles off the Humber and suffered heavy damage to the fore end. Having returned to Hull stern first and under tow, she was again rebuilt, thus demonstrating the sturdiness of trawler construction.

Provision of lifeboats had long been a problem in fishing vessels and in particular their launching, and the subject of many comments at Board of Trade inquiries. Completed in January 1957, the Grimsby steam trawler **Rodney** (1415), built without the aid of a WFA grant, had the distinction of being the first British trawler to dispense with lifeboats and carry inflatable life-rafts and a work boat under a Schat single arm davit. The **Rodney** was followed by the first of the "Cat" class for Grimsby owners. Best described as middle water trawlers, **Ross Tiger** (1416) was the first of the class of twelve vessels built by Cochrane and all had long and varied careers eventually ending up as offshore platform standby safety ships. **Ross Tiger** entered service in February 1957 under the command of G E Pedersen, known as 'Silent Jess' because of his aversion to using the radio. At 66 years of age, he was Grimsby's oldest working skipper. Son of a Norwegian fisherman he gained his skipper's ticket back in 1918 and in all that time **Ross Tiger** was his first new-build. The vessel still survives, preserved and open to the public at the Grimsby Fishing Heritage Centre.

In 1958 a new trawler hull form was developed by the Selby design team, with new hull lines and a transom stern, a model first being tested in the tank at the National Physical Laboratory at Teddington, then in Middlesex. The tests and subsequent adjustments resulted in a hull with more usable space, a 5% increase in speed and an equivalent reduction in propulsive power requirements. The first build to use this new design was the motor trawler **Kelvin** (1431) completed for Grimsby owners in September 1958.

*The **Kelvin**, described above.*

(Cochrane archive)

*A dramatic photograph illustrating the damage done to the **Loch Melfort** following the collision as described on page 6.*

(Author's collection)

The takeover by the Ross Group

In the late 1950s, the Selby shipyard was in dire need of cash injection to modernise and provide increased facilities as vessels continued to grow in size. The Ross Group Ltd of Grimsby was in expansionist mode at this time and was looking not only to acquire trawling companies, but also shipbuilding and repair facilities. Approaches were made to the Cochrane family who had built many of the ships for the companies that had come under Ross control, and the outcome was an agreement to buy the Selby shipyard, the sale being concluded on 21 September 1959 with Lewis Cochrane continuing as managing director of the company. In 1960 A Dalby was appointed assistant yard manager and a rolling programme of modernisation was created resulting in a spend of some £142,000, enabling the yard to build vessels of technically advanced design and increased size.

Taking full advantage of the yard improvements, the fishery research vessel *Clione* (1458) was launched in early 1961 for the Ministry of Agriculture, Fisheries & Food (MAFF) to be based at their laboratory in Lowestoft. Basically of conventional trawler hull form with transom stern, the bridge was set rather further forward and from there to the transom was plated to boat deck height to give increased accommodation for a laboratory, and cabins for five biologists/Lowestoft support staff. Specialist outfitting was carried out at Lowestoft and included laboratory fit-out, electrics for underwater cameras, freezer equipment and a tank to transport about 20 tons of live fish back to shore for examination. *Clione* made her first voyage later in the year to investigate herring stocks which included tagging of batches of fish. The vessel is believed to be still operating as the private ship, *Lynn G*. The first fishery research vessel built by Cochrane was the *Ernest Holt* (1342) completed in 1948.

In September 1963, seventy-seven years since the completion of the first steam trawler *Romulus* (11) at Beverley, Cochrane delivered their last side fishing trawler. The motor trawler *Stella Altair* (1485) and her sister *Stella Sirius* (1484) were to a well-proven design and it was fitting that Charles D Holmes & Co Ltd, Hull, who had manufactured so many of the triple expansion steam units for the Cochrane-built vessels should have provided the pair with the Werkspoor oil engine built by them under licence. With a very full order book including ice breaking tugs for Sweden and tugs for United Towing Ltd, Hull, and London owners, the shipyard was very busy. An order for a further tug to be built for Sweden was secured in 1964. The ocean going tug *Herakles* (1499), built with icebreaking capability, presented a real challenge to the shipyard. With cutaway forefoot and a very high weight per foot run on the short keel, great care was taken at the building berth and especially in the construction of the launch-ways.

The year 1963 saw seven launches and completions, the same output as the previous year, and saw orders placed for two versatile fishing vessels for Icelandic owners, the first from that country for eleven years. The *Lord Wakefield* (1118) although built in 1933 had electric welded shell plate butts and the two motor fishing vessels *Jorundur II* (1493) and *Jorundur III* (1494) completed in 1964, were the first all-welded

7

vessels built at the Selby shipyard. They were designed for three methods of catching, namely long lining, purse seining and trawling.

The tug **Hibernia** (1483) was delivered to William Watkins Ltd, the London tug owners, in January 1963. A sophisticated vessel intended for ship handling and coastal towage she had a powerful firefighting system with four fire monitors mounted on a platform 35 feet above the deck supported by a tripod mast. Each monitor could deliver 250 gallons of water or foam per minute and therefore capable of supporting firefighting efforts at both the Isle of Grain and Shell Haven oil refineries and other oil storage facilities on the Thames.

The last orders for boilers and steam reciprocating machinery had been placed with the long-established Hull boilermakers and engine builders, Amos & Smith Ltd in 1959 and now all machinery installations for Selby-built tonnage was subcontracted to Drypool Engineering Ltd, also in Hull.

For near water fishing, especially in the North Sea a new class of small motor stern trawlers was designed with unmanned engine room, bridge control and controllable pitch propeller. Working the gear from aft and with a long forecastle, the design also offered more safety for the crew but was somewhat vulnerable in a steep following sea. With this high level of automation the trawler could be operated and fished by a crew of only five and therefore offered great savings. The first to be completed was **Ross Daring** (1488) at a cost of £109,278 with a WFA grant of £23.700 and a further two vessels of the class followed but engine problems and unfamiliarity with the working of the gear resulted in them being sold in 1968. A fourth vessel, to have been named **Ross Duchess** (1502), was cancelled.

*The **Ross Daring** on trials.*

(Cochrane archive)

At the other end of the scale, Cochrane had secured an order from Ross Trawlers Ltd, Grimsby, to build a diesel-electric freezer stern trawler and it was to be, at 222.5 feet overall, the longest ship built at Selby so far. **Ross Valiant** (1489) was also the first freezer trawler to be built for Grimsby owners. She sailed on her maiden voyage on 24 June 1964 with Skipper

Jock Kerr in command and from a 35-day trip to the grounds off the Labrador coast, landed a capacity catch of 400 tons, equivalent to 6400 kits. This successful trip encouraged the Ross Group to go ahead with a class of freezer stern trawlers with even larger fish room capacity. The first of class was **Ross Vanguard** (1503) launched on 14 October 1965. It had been intended to build eight vessels but with the high cost of construction, in excess of £500,000 per vessel, only four were built.

*The **Ross Valiant** is assisted down the River Ouse by the **Lighterman** and **Pressman**.*

(Cochrane archive)

It is worthy of comment that in 1964 two early Cochrane-built vessels were still in the news. The **Albion of York** (1), the first vessel built in 1884 by Cochrane, Hamilton & Cooper at Grovehill Shipyard, Beverley, as a steam screw lighter, was still being used as a dumb lighter at Selby under the unofficially shortened name of **Albion**, and was not broken up until the 1990s. The steam trawler **Ulysses** (22), built at Beverley in 1889 and having been fitted with an oil engine as late as 1956, was only sold for breaking up at Stavanger in April 1964.

Although Lewis Cochrane retired as managing director in 1962 he remained as company chairman until his retirement on 31 December 1963. Sadly he died aged 73 almost two years later on 24 October 1965. His younger brother D M Cochrane who had started work with the company in the drawing office in 1915 had died on 15 October 1962 and S J Cochrane who had started work in 1925 had died on 20 April 1950 in his 58th year. Thus from the end of 1963 there was no longer any Cochrane family involved in the Selby shipyard.

Lewis Cochrane had gained valuable experience with the Grangemouth Dockyard Co Ltd and on his return to Selby in 1914 he became assistant manager and eventually yard manager. Elected to the board in 1917 during the war, he became managing director in 1929 and was chairman and managing director from 1943. The new managing director and ship director was D M Waite who was responsible for the development of a 'production flow system' using prefabrication techniques in the more spacious workshops. The installation of a 12.5 ton enabled bigger sections or weldments to be constructed in the sheds alongside the building berths and lifted into position.

John M T Ross was appointed chairman in succession to Lewis Cochrane on 13 January 1964 and thus became chairman of Cochrane & Sons Ltd and J S Doig (Grimsby) Ltd, the Grimsby shipbuilders and repairers as well as managing director of Ross Trawlers Ltd, Grimsby. On 17 December 1965, D B Cobb took over as managing director of Cochrane and in April 1966, W M George was appointed technical director. Later in the year Eric Hammal became chief designer and F F McGuire assumed the role of yard manager in mid-1967.

In the mid-1960s the London & Rochester Trading Co Ltd, Rochester, (L&RT) had seen an opportunity to invest in ships capable of supporting the growing exploration and exploitation of the North Sea gas and oil fields. In 1966 they sought tenders for the construction of an offshore platform/rig support ship and the Selby yard successfully secured the contract. **East Shore** (1505), completed in May 1966, was the first of a series of twelve ships the yard built for Offshore Marine Ltd, which after the delivery of **East Shore** became the trading name of the L&RT subsidiary. Initially following an improved design of the typical 'supply boats' used in the Gulf of Mexico oil fields in conjunction with oil exploration rigs, the design advanced quickly with experience gained in the North Sea. The series of offshore vessels and others ships ordered in the second half of the 1960s, were not all built by Cochrane at Selby as the company went through a transition period.

Drypool Group

During 1968 there was a proposed merger between the Associated Fisheries Group and the Ross Group.

The assets of both companies were valued for a new share issue for a new company, British United Trawlers Ltd. The trawler fleets were valued according to the cost of building, modifications, and as often, depreciation over a period of twenty years. These figures were entered as the book value in the new company accounts.

The new company did not want to be involved in shipbuilding and after some negotiations Cochrane became part of the Drypool Group Ltd, Hull, on 1 January 1969, with Bob Shepherd of R J Shepherd Engineering Ltd as chairman. Unfortunately Bob Shepherd died later that year and was replaced by Philip Curtis. Back in 1965, Drypool had taken over H J Shepherd Engineering of Bridlington, despite comments by Bob Shepherd to the press in 1963 that the prospects for the Drypool company were not good; they had a workforce of some 300 men at the time. The Drypool company had been in existence since 1915, but in 1921 the Rix family became involved, and as owners of a coasting fleet, used them for the repair of their own ships.

On 12 June 1968, **Rudderman** (1519), an oil products tanker, at 274 feet overall became the longest vessel to be launched at the Selby yard to date. Built for Helmsman Shipping Co Ltd a Rowbotham subsidiary, the Hull yard of Drypool Engineering & Drydock Co Ltd, having ventured into shipbuilding, had already built a series of tankers for Rowbothams and at the time had the tanker, **Leadsman** (Drypool yard no.23) on the stocks; by agreement she was launched at Hull the previous day. The Hull-built vessels were designed by Drypool but on the company acquiring the Selby yard, Cochrane drawing office became the design centre for the group.

The supply ship **Arctic Shore** (1524) intended for support work off Nova Scotia, was the first launch

*The **Arctic Shore** outbound from Great Yarmouth.*

(Stuart Emery collection)

under the new management and was named by Mrs K H Davison, wife of the exploration manager of Shell-UK Explorations & Production Co Ltd. With the launch of the small bulk carrier *Stirlingbrook* (1526) of 2900 tonnes deadweight for the European and Mediterranean trades on 12 September 1969, records were again broken, and at 283 feet overall she became the longest vessel launched to date. She was followed by the *Somersetbrook* (1536) in March 1970 of similar dimensions and a further three vessels followed.

The *Polar Shore* (1539), the first vessel built to metric measurements and completed at the end of 1971, was designed to operate in the harsh climate of Nova Scotia and further north in Newfoundland waters and was the world's first ice strengthened tug and supply vessel enabling her to operate in floe ice conditions. The deck machinery, windlass and winches were built by the Drypool Group subsidiary R J Shepherd Engineering Ltd, Bridlington. At the end of the year work began on the 6165 tonne deadweight tanker *Helmsman* (1540), which at 340.5 feet became the biggest vessel constructed to date, and the longest ever from the yard. At the launch on 5 November 1971, many spectators ignored warnings to stand fore and aft of the vessel as it entered the water and those watching from the opposite bank amidships were swamped with the huge water surge created. Four people were taken to hospital but fortunately there were no serious injuries. The incident was reported not only in the local papers but also in the national press.

Another large vessel was launched in October 1972, namely the 286.75 feet, 3202 tonne deadweight chemical tanker *Astraman* (1534). The ceremony was performed by Mrs Grace Thorne, wife of D Thorne, company secretary of C Rowbotham & Sons Ltd. The maximum size ship that could be accommodated at the Selby yard was estimated to be 344 feet with a maximum beam of 54 feet equating to a deadweight of some 7000 - 8000 tonnes.

The company which had been nominated as the shipbuilding division of the Drypool Group in 1970, became again Cochrane & Sons Ltd, in 1973. In 1972, Ken Marshall who had joined the company as an apprentice in 1959, became the production manager. Other changes in yard management around this time saw Bill George become technical director and Eric Hammal assume the role of naval architect, while later in August 1974, Norman Acaster joined as managing director responsible for shipbuilding following the departure of F F McGuire to join Dunstons at Hessle.

Receivership and formation of Cochrane Shipbuilders Ltd

From late 1974 the Drypool Group Ltd was experiencing financial problems and Norman Acaster, as managing director of Cochrane was heavily involved in keeping the Selby yard fully employed and maintaining the workforce. However, on 6 September 1975 the Drypool Group Ltd was placed in receivership. At this time the Group was involved in

Helmsman (1540)

(Cochrane archive)

shipbuilding and repair across eight sites. The number of employees is shown in brackets.

Selby shipyard (400)
Beverley shipyard (198)
Drydock & ship repair facility at Hull (205)
Alexandra Dock, Hull (159)
Maritime Electrics Ltd, Hull (45)
Shepherds Marine & Industrial Spares Ltd, Hull (10)
R J Shepherd Engineering Ltd, Bridlington (62)
Marine & Hydraulic Services Ltd, Hull (15)

Involvement with the Drypool Group had also meant that a number of Cochrane-designed ships were of necessity built in other yards or that Cochrane had only a marginal involvement. The **Maris 1** (Beverley yard no.1035) a patrol vessel for the Government of Nigeria, was launched at Beverley in August 1973 and completed in March 1974 after C D Holmes had closed down. The **Maris I** was a Beverley design, almost completed at the time the yard was purchased by Drypool and Cochrane were responsible only for final testing and sea trials. Three other vessels were launched at Beverley in 1974, **Seaforth Champion** (1559), **Stirling Rock** (1552) and **Tilstone Maid** (1562).

The supply ship **Stirling Sword** (1567) was built at the Drypool yard in Hull in 1976 and also from that yard came **Pacific Shore** (1523), **Island Shore** (1529), **Imperial Service** (1531), **Monarch Service** (1534), **Watercourse** (1545), **Waterfowl** (1546) and **Seaforth Hero** (1550).

The **Cape Shore** (1530), **Paramount Service** (1532) and **Majestic Service** (1535) were all built at the Charles Hill shipyard in Bristol, as were **Seaforth Prince** (1549) in 1973 and also **Seaforth Victor** built in 1975. They were all given Cochrane yard numbers as was **Bay Shore** (1538), built by J Bolson at Poole in 1971.

Very strong winds and gales were experienced over eastern England at the end of 1975 with wind speeds of up to 90mph. On 31 December the 75ft gantry crane in the yard was blown down and the following day the River Ouse flowed over the protective banks and flooded the yard.

On 1 June 1976, the company was bought from the Receiver of the Drypool Group Ltd by United Towing Co Ltd, Hull (which Basil Parkes had bought in 1960 and of which his nephew, Tony Wilbraham, was now chairman), with a promised immediate investment of £500,000 and the prospect of future employment for 70 more men, the workforce having been reduced to 373 employees since Drypool was placed in receivership. United Towing had taken advantage of a loan of £400,000 under the normal Industry Act terms to ensure that sufficient capital was available for the yard.

The new chairman was A (Tony) D Wilbraham; the managing director Norman Acaster; technical director William M George; company secretary R C Nelson and Ken Marshall the production manager. Shortly after Cochrane Shipbuilders Ltd was formed as a subsidiary of United Towing Co Ltd, the last two orders taken by Drypool, **Seaforth Conqueror** (1573) and **Seaforth Clansman** (1574) were completed by this company. New yard numbers starting at 101 were then introduced.

*A fine view of the **Seaforth Conqueror** (1573) as she was fitting out.*

(Cochrane archive)

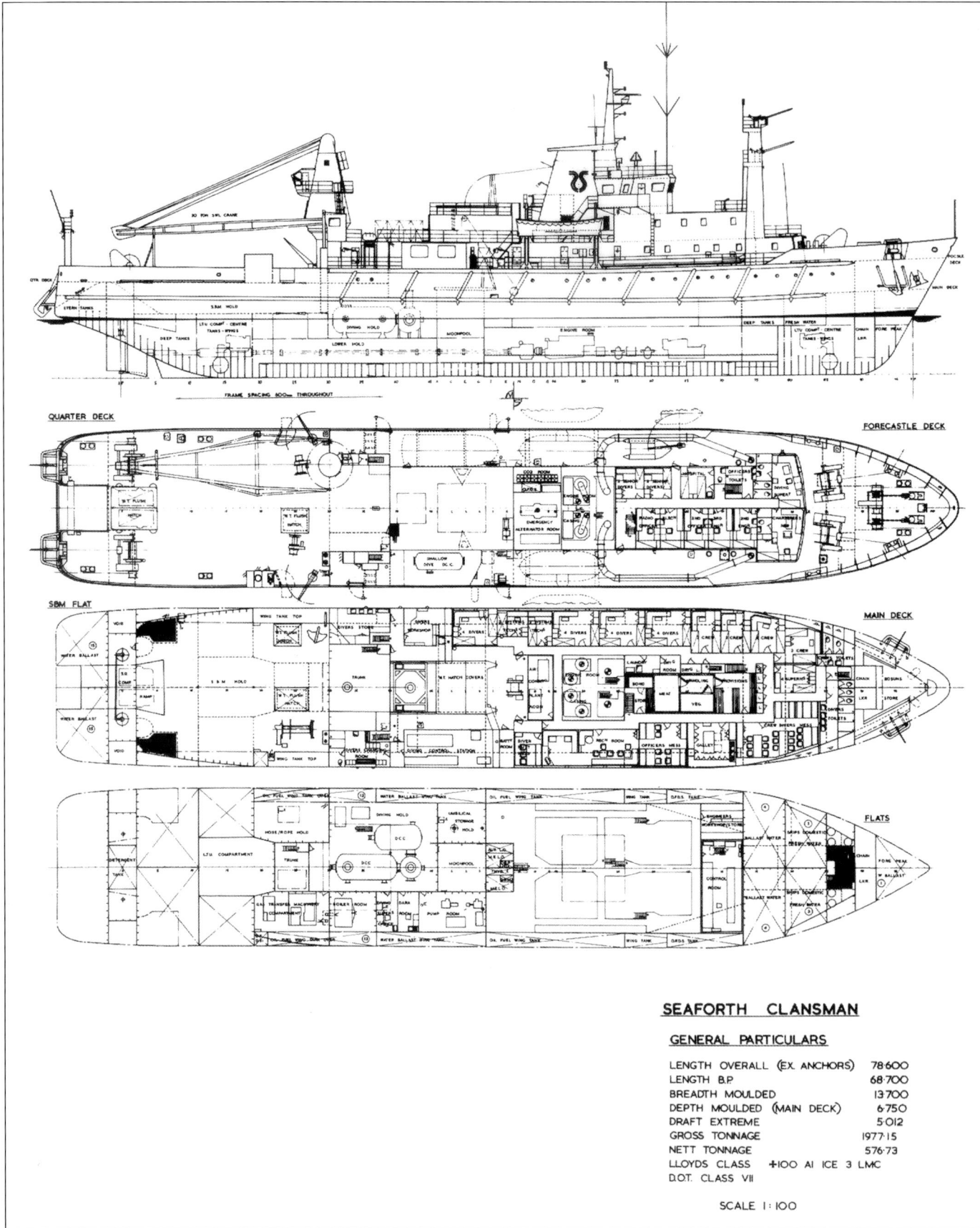

SEAFORTH CLANSMAN

GENERAL PARTICULARS

LENGTH OVERALL (EX. ANCHORS)	78·600
LENGTH B.P.	68·700
BREADTH MOULDED	13·700
DEPTH MOULDED (MAIN DECK)	6·750
DRAFT EXTREME	5·012
GROSS TONNAGE	1977·15
NETT TONNAGE	576·73
LLOYDS CLASS	+100 AI ICE 3 LMC
D.O.T. CLASS VII	

SCALE 1 : 100

*General arrangement drawing of the **Seaforth Clansman**.*

(Cochrane archive)

The £2.5 million offshore supply vessel tug and anchor handler **Seaforth Conqueror** was launched in June 1976. Her engine room layout comprised four Mirrlees Blackstone main engines delivering 7320bhp, geared to two shafts with controllable pitch propellers in fixed nozzles, but with the ability to use one engine on each shaft. Before the launch, George Wiseman (1905-1995) was commissioned by Seaforth Marine to paint a picture of her as she would appear when working in the North Sea with an oil platform close by. Born in Aberdeen, George P Wiseman had been painting Cochrane vessels since 1956 and many of his works were hung in the boardroom or were used to illustrate the company calendars. Harry Hudson Rodmell (1896-1984), marine and graphic artist, provided designs in the 1930s and 1940s for magazine advertisements and company brochures.

Although **Seaforth Clansman** (1574) had been ordered and given a yard number under the old sequence, she was the first ship launched under the newly-formed Cochrane Shipbuilders Ltd. Intitially ordered as a sister ship to **Seaforth Conqueror** she was rebuilt and redesigned whilst under construction, with a new 10 metre section inserted amidships. She was completed as a diving support vessel but also with firefighting and pollution control capability. With accommodation for 46 men including 25 divers, the saturation diving system and 30-ton crane made her an extremely versatile vessel. In 1981 she was chartered by the Ministry of Defence (Navy), London, to accommodate Naval Party 1007 as a deep sea saturation diving platform following the decommissioning of HMS **Reclaim**. She was used by Naval Party 1007 in March 1982 to explore the wreck of Britain's first submarine, **Holland I**, to assess the possibility of her being raised for preservation ashore.

Evolution of the Selby shipyard and some notable ships

From the signing of a contract to completion, trials and handing over, was a period of between 7 and 18 months depending on the size and complexity of the vessel. At any one time there was usually a newly laid keel, one vessel nearing launch and one vessel alongside fitting out. The building berths were alongside and parallel to the River Ouse and the ships were launched broadside into the water at slack water on a high tide. Depending on size, one or two tugs would be in attendance and once in the water the vessel would be laid alongside the fitting out quay for further work to be undertaken to completion.

The maximum length of vessel that could be built at Selby had increased over the years but was now finalised at about 105 metres (344 feet), the limiting factor not being the shipyard layout but a series of tight bends on the River Ouse between Selby and Goole. The M62 road bridge at Goole was now also a limiting factor with only about 25 metres clearance.

The breakdown of construction into units or weldments was by now well established in the yard, these units being prefabricated in the sheds up to a weight of 15 tonnes. They were then taken to the riverside building site where the complete vessel was assembled. Three to six vessels were built each year using approximately 2500 tonnes of steel. All fitting out was carried out at the shipyard and with vessels now engined before launch it only remained to check shaft alignment when afloat before sailing for Hull under their own power for drydocking and final paint. Acceptance trials were held in the Humber and on the Sunk Island measured mile.

Through the 1950s and 1960s based on the experience when building the two trawlers (1309) and (1319), a basic contract price was established for hull and propulsion machinery with various clauses allowing for increases in the cost of steel, and of shipyard and engineering wages. Such contracts were increasingly replaced with fixed price contracts which attracted heavy penalties for late delivery, or failure to meet design specifications, particularly speed, fuel consumption and in the case of cargo ships, cargo capacity. Staged payments were made as agreed between the shipyard and the ship owner with a 5% payment upon handover of the vessel.

If the yard made use of the Government Intervention Fund, then the price was fixed and the company was not allowed to make any profit on the contract. If in fact a profit was made then that sum had to be paid back into the Fund. If on the other hand the yard had miscalculated and made a loss, that had to be stood by the company and could not be reclaimed.

The post-war years had seen the Selby yard output dominated by trawler building, but now a large variety of ships were being constructed including dry cargo vessels, offshore supply vessels, tankers of different types, tugs, again of different types, ferries and one research vessel as well as pioneering the design and build of stern trawlers, including freezers and a variety of side fishing trawlers. The design department and drawing office at Selby also prepared designs for vessels to be built at other shipyards. One such ship was the **Salvageman** for United Towing, built by Chung Wah Shipbuilding & Engineering Co Ltd, Hong Kong (160). At the time of her construction in 1980, she was the largest and most powerful tug on the British register.

A standard design for a twin screw harbour tug developing 2920bhp was created jointly by Bernard Ingram, managing director of Humber Tugs, and Bill George, technical director of the shipyard, with an estimated cost to build of £1 million. The first of a pair, **Lady Moira** (101), was also the first vessel in the new numbering sequence, and was launched in April 1977 by Mrs Moira Smith wife of the Official Receiver appointed after the collapse of the Drypool Group.

A notable ship built in the late 1970s was the **Selbydyke** (106) built for the Klondyke Shipping Co Ltd, Anlaby, Hull. At 79,38 metres overall (260 feet) she was the largest of her type to be built at the yard until that time. With a long single hold designed to carry timber, pipes, steel and indivisible cargoes, she was propelled by a 43 ton Mirrlees Blackstone six cylinder engine, the biggest ever fitted to a vessel at the yard.

She was launched on 16 November 1978 by Mrs Ruth Atley, wife of the chairman of timber importers May & Hassall Ltd and left Hull the following year under the command of Captain H Upperton, the longest serving master of Klondyke Shipping.

On 31 December 1978 the North British Maritime Group Ltd, Yarm, was formed and amongst other companies owned by the Parkes family and others, United Towing Ltd became a subsidiary as did Cochrane Shipbuilders Ltd. The production manager Ken Marshall was designated production director and became a member of the board of Cochrane Shipbuilders Ltd.

The **Selbydyke** was followed by the 38,26 metre (125.5 feet) diving support ship **Star Perseus** (107) for Star Offshore Services Marine Ltd, Aberdeen. Built at a cost of £1.5 million she had accommodation for fourteen divers and crew and was initially deployed in North Sea support. In 1988 She was sold to the Government of New Zealand (Ministry of Defence), Wellington for NZ$1.5 million. On arrival in New Zealand she was converted for diving and mine counter measures support and renamed HMNZS **Manawanui**. She remains in service but it is intended to replace her by 2018.

Costing more than £1 million, the dry cargo vessel **Angelonia** (110) was the second of two sisterships for J Wharton (Shipping) Ltd, Gunness, and was launched on 19 February 1980 by Mrs Fiona Lisbeth Wharton, wife of the managing director, John Steven Wharton. The ship was completed a month ahead of schedule. Now fitted with a deck excavator she remains in service in the Faroe Islands as the Runavik-registered **Hav Tind**.

The **Esso Plymouth** (111), launched in June 1980 by Dr Patricia Pearce, wife of Esso Petroleum's retired chairman, Dr Austin Pearce, was the first UK built ship to be fitted with a resiliently mounted deck housing. Designed as a clean products tanker she was capable of carrying light petroleum products, kerosene, petrol, avtur (ATF) and diesel.

In 1981 the tug **Lady Elizabeth** (114), the first of two 'tractor tugs' for Humber Tugs Ltd, was the first vessel the yard had built fitted with Voith Schneider cycloidal propeller units. These units combine propulsion and steering, resulting in a highly manouevrable tug. She was launched on 22 April 1981 by Mrs Margaret Tebbitt, accompanied by her husband Norman Tebbitt,

*The huge Mirrlees engine being lifted carefully into the **Selbydyke**.*

(Cochrane archive)

Industry Minister in the Conservative government of the day.

On 2 April 1982, Argentinian troops invaded the Falkland Islands and South Atlantic territories and on 5 April the British government dispatched a naval task force to engage and retake the Islands. Selby-built tugs were part of the STUFT (Ships Taken Up From Trade) ships that accompanied the task force or supplemented it after retaking the Islands on 14 June 1982. The **Irishman** (103), **Yorkshireman** (104) and **Salvageman** (built in Hong Kong but designed at Selby) all joined the naval operation.

*The **Irishman** after launch with the Selby-built **Waterman** passing.*

(John Adamson)

An interesting departure from the normal run of tugs and offshore supply vessels was an order placed by Land & Marine Engineering Ltd, Bromborough, a subsidiary of Royal Boskalis Westminster BV, Papendrecht, for three twin-screw firefighting/tug/mooring tenders. Intended for offshore terminals, **Lanmar Crest** (116) was the first to be launched on 6 July 1982. All three eventually worked at the Forcados oil terminal in the Niger delta, Nigeria.

When the naval task force sailed on 5 April 1982 to retake the South Atlantic islands, Admiral Sir John Fieldhouse, as C-in-C Fleet at Northwood, was in overall command. On 26 November 1984 the tug **Seaman** (129) was launched by Lady Margaret Fieldhouse, wife of Sir John, now the First Sea Lord who was also present. The tugs of United Towing had been prominent in the Falklands campaign, and two years after the end of hostilities their vessels were still being employed by the Ministry of Defence in and around the Falkland Islands and South Georgia.

*The champagne is about to be released at the launch of the **Seaman**.*

(John Adamson)

The Falklands connection was maintained with the cargo vessel **Norbrit Faith** (121) completed in December 1982. Acceptance and final preparations were made in Hull before sailing for Marchwood, Southampton Water, to load and sail on her maiden voyage to the South Atlantic. The first of two vessels for North British Shipping Ltd, Hull, (part of the North British Maritime Group, owners of the Selby yard) and of an advanced design with a single box hold, they were capable of self discharge via a track mounted gantry crane. Because of the limited port facilities at that time in the Falklands, a self contained vessel of this type was ideal. The **Norbrit Hope** (122) was launched in November 1982 by Mrs Debbie Wilbraham, a former beauty queen and wife of David Wilbraham, managing director of North British Shipping.

By the end of 1982, the fishing fleet of Boston Deep Sea Fisheries Ltd, now managed by Neil Parkes, had been reduced to three trawlers, with twenty-seven trawlers converted to offshore platform safety ships. Tony Wilbraham was a director of Star Offshore Services, a company engaged in offshore support/supply/safety vessels and an amicable agreement was reached in 1983 for the purchase of the Boston company. Thus the Wilbraham family was now in sole control of Boston Deep Sea Fisheries.

The sister ships **Stridence** (123) and **Turbulence** (124), 1800 tonnes deadweight single screw river sea ships, were launched for Crescent Shipping in March and July 1983 respectively. Designed for bulk cargo and strengthened to carry steel coils, they also had a retractable wheelhouse for up river passage.

They were followed in October 1983 by the **Willonia** (125), 2415 tonnes deadweight and the first of three near sisters, two for J Wharton (Shipping) Ltd and one for lease to F T Everard & Sons Ltd, Greenhithe, with whom Whartons were associated. With a single box hold and portable bulkheads for the carriage of grain, each of the trio was powered by a 6-cylinder MAK oil engine developing 1285bhp driving a controllable pitch propeller. **Willonia**, named after the two year old son of managing director, Steve Wharton, had the distinction of being built almost entirely of Scunthorpe steel. The second of the trio **Selectivity** (126) was completed three weeks ahead of schedule and on her maiden voyage loaded grain at New Holland for the Continent.

The early 1980s were difficult times for British shipbuilders with intense competition from overseas yards. On 1 September 1977, in an attempt to secure orders and maintain the workforce, the majority of British shipyards had been nationalised by the Labour government to form the British Shipbuilders Corporation with headquarters in Newcastle. The Corporation initially included 32 shipyards, 6 marine engine works and 6 general engineering plants; Cochrane Shipbuilders with a reasonable order book was not nationalised. By 1982 British Shipbuilders had closed half the shipyards and several others were in serious trouble. One of those yards was Goole Shipbuilders Ltd, established in 1901 as the Goole Shipbuilding & Repairing Co Ltd, which in November 1977, had been placed in the Swan Hunter Group - Small Ship Division. Renamed Goole Shipbuilders Ltd on nationalisation, the yard at Goole was closed in April 1984 with the loss of 365 jobs.

At Selby, Cochrane continued to hold a satisfactory order book and at the end of 1984 it was decided to buy the 11 acre shipyard at Goole. In 1985 the yard was reactivated and a workforce of 250 recruited, though with some interchange of labour between the two yards. The first ship built at Goole was the offshore supply vessel **Stirling Esk** (132) for Harrisons Clyde Ltd to be followed by **Wimpey Seawitch** (134); the Scarborough white fish trawler **Allegiance S** (142), which had been transferred from Steelships Ltd, Truro, when their parent company went into receivership; three 24,38 metre (80 feet) flat topped open barges (145-147) and finally **Lady Sybil** (138) which was completed in December 1987.

The **Selectivity** at Dundalk on 26 July 1994. She had arrived from Calais and loaded grain for Rotterdam.

(Charley McCarthy)

In April 1985 following a fierce competitive tendering exercise, Caledonian MacBrayne Ltd, Glasgow, placed an order for a £5.5 million ro/ro cargo passenger ferry for service in the Western Isles; this was the first MacBrayne vessel to be built outside Scotland. The twin screw, 15 knot, **Hebridean Isles** (130) was designed to handle 80 cars or 10 commercial vehicles and 500 passengers during the summer schedules. At the time she was the most expensive single vessel built by the yard and at 15,8 metres (51.8 feet) became the widest built so far. The launch at Selby on 4 July 1985 by HRH the Duchess of Kent was a splendid occasion, with the Glasgow Skye Pipe Band in attendance and many dignitories from both sides of the border. The shipyard was decorated with flags and packed with almost 3000 people, VIPs, guests, employees and their families and a very large number of flag-waving local school children. Lunch was taken in a marquee erected in the shipyard and the Duchess was presented with a large Edinburgh crystal rose bowl engraved with the Yorkshire rose and the Scottish thistle, especially commissioned for the occasion.

Staying with ro/ro cargo passenger ferries, a £5 million order was placed in 1986 by the Eurolease Corporation Ltd, London, for a ferry for charter to Sealink Isle of Wight Ferries for their Portsmouth to Fishbourne route. Launched on 4 November 1986, **St Cecilia** (135) was designed to carry 142 cars and 1000 passengers, and at 16,8 metres (55.1 feet) beam she became the widest vessel ever constructed at the yard.

On 30 December 1986 a £5 million order was placed by J Marr & Son Ltd, Hull, with the Selby yard to build two wet fish stern trawlers of an advanced design. This was the first deep sea fishing vessel order to be placed by the Marr company in twelve years. The **Thornella** (143) was launched on 21 December 1987 by Mrs Bridget Marr wife of Charles Marr and sister ship **Lancella** (143) on 18 April 1988 by Sophie Marr, daughter of Alan Marr who was chairman of the Marr company. Although James Herbert Marr had chartered the first Cochrane-built steam trawler **Romulus** (11) in 1911 and in the same year J Marr & Son Ltd, Fleetwood, placed their order for the **Velia** (536), their first order with Cochranes.

In 1986 orders became difficult to obtain and there were redundancies at both Selby and Goole leading to the Goole yard being finally closed in 1988 following completion in January of the **Lady Sybil** (138). At Selby Norman Acaster retired as managing director of Cochrane and was replaced by Ken Marshall. Bill Edgar, a graduate of Birmingham and Strathclyde Universities was appointed executive chairman of Cochrane Shipbuilders. At this time the yard employed some 350 workers, loftsmen, shipwrights, boilermakers, platers, joiners, fitters, plumbers, welders, caulkers and electricians. Although there had been many incomers in the managerial and technical roles, most of the work force were local people. Like many other shipyards in Britain, where the inherited skill of the work people was exploited to the full, there was a strong family element and in many cases several generations had worked at the yard, some with connections with the setting up of the company at Selby in 1901 following the move from Beverley.

Purchase by Howard Smith

In 1987, Howard Smith Pty, Sydney, Australia, a company with interests from sugar to shipping, bought a 75.1% interest in the North British Maritime Group for the sum of Aus$12 million, with the Wilbraham family retaining the remaining 24.9% stake. This allowed the Australian company to take over the tugs owned by the Group and to acquire Cochrane Shipbuilders Ltd.

The Selby yard had not been nationalised in 1977, but in order to compete with the nationalised yards which were receiving large sums of money from public funding, the yard management had to be very focused. Without the burden of a multilayered management structure, they were able to compete for orders and this was one of the factors that the new owners took on board when deciding to back the company in continuing to build ships. Ken Marshall, with twenty-seven years experience from apprentice to board room, remained as managing director of the shipyard and workforce which numbered a little under 400.

Although the workforce was unionised, disputes were usually settled fairly quickly on a direct basis between management and production. Demarcation between various trades, which had been the bane of British shipyards, was much reduced in the 1980s and a degree of interchangeability was agreed. It was possible then to set up a bonus scheme. If the man hours allocated on a particular ship, or even some tasks, were completed inside the allotted hours in the original cost estimate, then a bonus could be paid.

While Eric Hammal was naval architect, the company had installed computers in the design offices but these were largely used for calculations and not linked directly to production. Data was transferred to the drawing office and all working drawings produced by experienced draughtsmen. In later years, the mould loft had produced information for several ships by drawing the plates on screen, taking maximum advantage of the plate area and linking the computer to the installed plasma-arc cutting machines.

The next stage would have been for the drawing office to use CAD (computer-aided design) software, incorporating most of the information that the loft was preparing for production. However, capital investment in new plant and systems was always limited under the various ownerships.

By 1988, British shipbuilding was in a very depressed state, unable to compete with labour costs in many overseas countries and even the best-run yards like Selby were affected. One of the factors was the Icelandic 'Cod Wars' over the 200 mile limit around the island, which had seen the decline and almost wiping out of the distant water trawler fleets of Hull, Grimsby and Fleetwood. The other specialist trawler building yard in England, that at Beverley, had finally closed in 1975.

There were 70 redundancies at Selby in 1988 and Howard Smith put the yard up for sale, but at that stage they were simply approaching the market to see if there was any interest generated. The last of four tugs in a £4 million order for Humber Tugs, *Lady Dulcie* (139), which had been laid down as *Lady Thelma*, was launched in June 1988 by Mrs Dulcie Atkins, wife of Robert Atkins, Minister for Industry and Aerospace in the Conservative government.

Managing Director Norman Acaster (on left) presents a retirement gift of a portable television to joiner Alf Adamson on 9 November 1981 following Alf's 48 years service at the yard.

(John Adamson)

In November of 1988 a £2.5 million order was secured to build a small ro/ro vehicle and passenger ferry for charter to Orkney Ferries Ltd, Kirkwall. The vessel *Varagen* (162) was completed in June 1989 and placed on the Gills Bay (four miles west of John O'Groats) to Burwick, Orkney, service and with a speed of 14.5 knots, the passage time across the Pentland Firth was reduced.

A splendid aerial view of Varagen.

(Cochrane archive)

An ambitious order to build six schooner rigged luxury sailing charter 'yachts' was taken in 1990. Designed by Laurent Giles Architects Ltd, Lymington, and based on a Turkish 'gulet', the first and only one completed, *Danielle of Hamm* (156) was launched in September and lifted into the River Ouse by crane. The remaining yachts in the order were never built, resulting in a heavy financial loss to the company.

Ironically Cochrane had a full order book worth some £33 million, including a £13 million order from the Union Transport Group to build a series of low air draught cargo ships. With a depleted workforce and a tight delivery schedule two of the hulls of *Bromley Pearl* (164) and *Union Saturn* (165) were subcontracted to Richard Dunston (Hessle) Ltd. The design called for 3200 tonne deadweight ship with a single box hold, a large flap rudder, bow thruster, a telescopic wheelhouse lowering system, and as river sea ships they were colloquially known as 'Rhine traders'.

The Union Transport order ended in disaster. The hatch covers of *Union Jupiter* (163) and *Bromley Pearl* presented technical problems, and because of delays the contracts were cancelled for *Union Saturn* (165) and *Union Mercury* (166). Both vessels were subsequently sold on to other companies and continue to trade successfully. *Superiority* (167) launched in December 1990 for an Everard subsidiary was of the same hull design but with a different fit-out including a free fall lifeboat. She was sold to the Union Transport Group in 2002 and became the *Union Emerald*.

The final fishing vessel built by Cochrane was the wet fish trawler *Glenrose I* (148), for the Onward Fishing Co Ltd, Hull. She was launched in June 1990 by Miss June Taylor daughter of the company's managing director and skipper, Terry Taylor. As the Latvian *Glenrose* (LVV1555) the trawler is still fishing.

The oil product tanker *Eliza PG* (171) which cost £7.5 million and her sister ship, *Forth Bridge* (173) at almost 7500 tonnes displacement were amongst the heaviest vessels built at the yard. The *Helmsman* (1540), however, at 103,8 metres (340.5 feet) and displacing 7900 tons, was actually the heaviest and longest ship built at the Selby shipyard.

The last vessel completed at Selby since the foundation of the yard, ninety years previously, was the oil products tanker *Forth Bridge* (which would have been 1647 under the old numerical series). She left Selby on 13 September 1992 for dry docking and final paint at Hull and in the best traditions of the company was handed over to Forth Tankers plc on the contract date.

With limited plant and workers, the final two or three years before closure had seen large sections of steel work (the largest transportable by road) subcontracted to other yards including Swan Hunter on the Tyne. There were further redundancies in 1992 and a 37-hour week was introduced. Howard Smith then withdrew guarantees required for any new ship contracts and the yard was offered for sale in earnest. There was no interest and no offers were received for the yard as a going concern and the site was sold for development.

The entire contents of the shipyard were auctioned off in March 1993, including a number of ship models, but Ken Marshall the managing director, ensured that the archives and many ship drawings found a safe home in Hull Maritime Museum. Subsequently these have been transferred to North Yorkshire County Record Office at Northallerton.

Among the many historic artefacts catalogued there was a builder's model of the Grimsby registered (GY355), Fleetwood steam trawler *Margaret Rose* (1100) built in 1931 for Boston Deep Sea Fishing & Ice Co Ltd, Fleetwood, and sold to French associates in 1933. The model was bought by Colin Dent of Dent Steels Services Ltd, Bradford, suppliers of the majority of steel plate and sections to Cochrane for some twenty years prior to closure.

Gilbert I Mayes

Written with the aid of material researched and recorded by Arthur G Credland, assisted by the late Ken Marshall, Eric Hammal and Mike Thompson.

Eliza PG on trials

CHRONOLOGY

1941   J Press appointed foreman fitter in place of E Soulsby

1943   J McGill (foreman riveter) received BEM (probably a kinsman of the Cochrane's, Andrew Snr. had married Jane McGill)

       Thomas McGill Cochrane died 31 May

1945   J Triffet appointed foreman labourer in place of G Beckwith

1947   Eight vessels delivered for Icelandic government

1948   **Ernest Holt** (1342) - fisheries research vessel

       A L Cochrane awarded Grand Cross of the Order of the Falcon, Iceland

       Robert J Coles commenced as assistant yard manager on 4 October

1949   H Wright commenced as cashier

       E Marshall appointed foreman painter in place of R Braithwaite

1951   D M Cochrane resigned from Selby UDC

       A L Cochrane president of the Shipbuilding Employers' Federation

       A Clifford Kelly commenced as chief draughtsman

       R W Walker appointed assistant chief draughtsman

1953   A L Cochrane received OBE

       Victor Gray, naval architect, received MBE

1954   J McGill retired March, died during September

       Directors: A L Cochrane, OBE, 'Ouseburn', Doncaster Road, Selby

                      D M Cochrane, 'Rudyard', Green Lane, Selby

                      S J Cochrane, 59 Doncaster Road, Selby

                      V Gray, MBE, 45 Doncaster Road, Selby

       Secretary: D M Cochrane; naval architect V Gray; chief draughtsman A C Kelly.

       (Number of berths in yard 6; capacity up to 200 feet; maximum output 15,000 - 20,000 tons)

1955   **Arctic Warrior** (1367) - first winner of the Silver Cod trophy

       V Gray, naval architect, retired 1 July

       A Clifford Kelly appointed naval architect

       J Sheridan retired as foreman plater

       R Skinner appointed foreman plater

1956 J Freeman appointed head of time office; J Parks retired
1957 Frank Whitehead received BEM
1958 *Kelvin* (1431) - first trawler with transom stern
N Simms appointed foreman blacksmith; J Cantle retired
R Simm appointed steel stock clerk; H Gilleard retired, died March 1960
1959 Trawler *Corena* (1439) - first CP propeller
Cochrane became part of Ross Group of Grimsby in September
T Oxley appointed foreman joiner
F Cass retired; died July 1960
H Wright died August
R C Nelson appointed cashier
A Hardy appointed foreman welder; J Borland retired
1960 A Dalby appointed assistant yard manager
Sidney John Cochrane in hospital on 7 April; died on 20 April 1960
W Lodge appointed foreman shipwright; H Slack retired
E Sowerby retired
G Young appointed foreman riveter; E Middleton retired
A Clifford Kelly on 'Patton' Committeee; appointed director 1 June
R W Walker appointed acting chief draughtsman
1961 *Clione* (1458) - fisheries research vessel
Douglas M Waite appointed General Manager 1 May
H Bulling retired as foreman loftsman; started in 1899; succeeded by his son Harry Bulling as
    loftsman
1962 D M Cochrane to hospital on 14 June; managing director from 1 August; died on 15 October
Douglas M Waite appointed director in July
R C Nelson appointed company secretary in July
V Richards appointed cashier in July
A L Cochrane retired as MD end of July; remained chairman
W M George appointed assistant naval architect; then naval architect in October
J Press retired as foreman fitter end of December
1963 *Ross Daring* (1488) - first stern trawler
Jorundur II (1493) - first all-welded vessel
R McLauchlan appointed foreman fitter
Douglas M Waite appointed MD on 11 February
I Ingleson received BEM in June
W A Smith appointed assistant chief draughtsman in July
Andrew Lewis Cochrane retired end of December
All fitting out of vessels in Hull (Princes Dock) subcontracted to Drypool; no Cochrane employees
    in Hull
1964 John M Ross appointed director and chairman on 13 January
Robert J Coles appointed general manager
H Sykes retired in September after 60 years service
A Clifford Kelly retired in November
1965 *Ross Valiant* (1489) - first diesel-electric stern trawler
D B Cobb appointed Director
Douglas M Waite transferred to Ross Group, Grimsby
W Lodge died May; Tony Elliott appointed foreman shipwright
J Freeman left in September to join Shipbuilding Employers' Federation
V Gray died in September
Andrew Lewis Cochrane died in October
A J Dalby transferred to Ross Group, Grimsby, in October
F F McGuire appointed assistant shipyard manager in December
D B Cobb appointed MD in December
R W Walker took over steel purchasing
1966 *East Shore* (1505) - first offshore supply vessel
A J Dalby appointed chief estimator and planner; left in November
W M George appointed technical director in April
W A Smith appointed chief draughtsman in November
Douglas M Waite left in December to join Brooke Marine
Eric Hammal appointed chief designer
1967 T H Foster appointed chief estimator in January
K W Marshall in charge of planning in January

F F McGuire appointed yard manager in August

R Skinner, steelwork manager, died in October

B Walker and I G Johnson, technical assistants, and then to naval architects' department in November

1968   I Ingleson retired February

D B Cobb appointed MD of Ross International in April

R J Coles appointed director in June

W Will appointed outfitting superintendent in Hull, October; left December

J T Marson appointed steelwork section leader, October

R W Walker retired during October

1969   *Pacific Shore* (1523) - first vessel subcontracted to Hull, No.2 Dry Dock

Cochrane became part of Drypool Group

Directors: R J Shepherd, P Curtis, C Shepherd, W M George, D Townend

Engineering draughtsmen move to Selby

E Coultas, chief engineering draughtsman

T Hughes appointed storekeeper; R Chatterton retired

R J Shepherd died of heart attack in April

P Curtis appointed MD

R C Nelson left for Brooke Marine

1970   *Cape Shore* (1530) - first vessel subcontracted to Charles Hill, Bristol

Cochrane became part of Drypool Group Shipbuilding Division

1971   *Bay Shore* (1538) - first vessel subcontracted to Bolsons of Poole

G Guest appointed technical manager; from Cunard International Technical Services

1972   *Helmsman* (1540) - largest vessel ever built at yard

F F McGuire appointed production director

Ken Marshall appointed production manager

1973   *Astraman* (1543) - first chemical tanker

Selby reverted to Cochrane and Sons again

Eric Hammal appointed naval architect

C D Holmes, Beverley, shipyard acquired

1974   *Stirling Rock* (1552) - first vessel subcontracted to Beverley

F F McGuire left for Dunstons, July

G Guest left for Zapata Offshore

T Highlands appointed technical manager

N Acaster joined as shipbuilding managing director, August

1975   Drypool Group in receivership on 6 September; Robert C Smith appointed Receiver

Order book of fourteen vessels when in receivership, all to be completed except one tug for Cory transferred to Richards at Lowestoft

P Curtis and D Townsend resigned; N Acaster responsible for Group to the Receiver

E Marshall retired; P Knight appointed chargehand painter

1976   All parts of Drypool Group sold or closed

Selby Yard established as a stand-alone company. Cochrane Shipbuilders Ltd, became part of United Towing Co Ltd on 1 June; purchased from Receiver. No redundancies, continuity of contract conditions

Directors: A B Wilbraham, Sir Basil Parkes, F Stubbs, N Acaster, W M George, S Bashford

E J Wright appointed engineering technical manager

R C Nelson appointed chief accountant / company secretary

H M McCollin appointed electrical technical manager

C Smith appointed personnel, training and safety officer

T Elliot appointed shipyard manager

1977   *Seaforth Conqueror* (1574) - first diving support vessel

W Wood appointed assistant engineering manager to L Cooper

T Wright appointed foreman fitter

1978   Ken Marshall appointed production director on 31 January

1979   *Boston Sea Lance* (108) - first fully refrigerated cargo vessel

K Elliot appointed foreman joiner; T Oxley retired

M Fierheller appointed secretary to MD

B Walker appointed quality control and planning manager

T H Foster retired

P N Wilbraham appointed director

G Young retired as berth foreman

1982   *Lady Elizabeth* (114) - first Voith water tractor

1983	K Jakowyszyn appointed head loftsman; H Bulling retired
	P Hainsworth appointed data processing manager
	W A Smith died of heart attack
	R C Nelson appointed director
1984	Goole Shipyard acquired from British Shipbuilders; N Woffendon appointed Goole shipyard manager
1985	*Hebridean Isles* (130) - first vehicle/passenger ferry
	Stirling Esk (132) - first new building order for Goole shipyard
	P Jackson retired as maintenance foreman
1986	J McCreadie appointed repair director at Goole
1987	E Howden, diving subcontractor from Selby, died under stern of *St Cecilia*
	N Acaster retired
	Ken Marshall appointed MD
	Australian Company Howard Smith acquired 75% of North British Maritime Group
	T Elliot, yard manager, died of heart attack May
	K Moss appointed director
	E J Wright returned and appointed sales director
	J Shannon appointed production manager
	A Jones appointed sales manager
	C Smith retired
	R McLachlan retired
	Goole Shipyard closed, labour transferred to Selby
1988	R C Nelson resigned in January
	F W Mart appointed finance director in January; then company secretary
	P Love appointed company accountant
	W M George took early retirement
	Wilbraham family removed from board by Howard Smith in July
	W Edgar appointed executive chairman on 3-year contract
1989	R Prestwood appointed systems information manager
	J Marson had heart attack and died 17 November
1990	*Danielle of Hamm* (157) - first vessel to be launched by crane
	E J Wright left in April; returned in May
	W Edgar left end of May
	Howard Smith own 100% of North British Maritime
1991	M Lewis appointed production director in February; left in July
1992	*Forth Bridge* (174); last vessel, delivered end of October
1992	Howard Smith announced no further support for Cochrane on 12 August
1993	All yard and office equipment sold at 3 day auction 23, 25 and 26 March

Industrial Relations /Organisation/Yard Estate

1942	Second canteen opened
1943	Furnaces now converted to burn creosote pitch
1944	PAYE commenced
1946	Paid holidays increased to 6 days annually
1947	Yard flooded along with most of Selby and district
	44 hour working week commenced
1948	National Health Scheme commenced
1949	Sports club commenced
1951	Paid holidays increased to 2 weeks annually
	First staff party at Monk Fryston Hall
1954	Yard flooded but not seriously
1957	Hydraulic plant converted to Towlers system
	Steam plant scrapped and chimney felled
1958	Second staff pension scheme introduced
	No.2 Jetty crane electrified
1959	No.1 Jetty crane electrified
1960	Yard modernisation commenced
	New yard management offices completed
	42 hour working week commenced 28 March
	New flood bank built up and completed in July
	Main office extension completed on 20 September
	New sub-station on river bank commissioned for welding circuits

1961　Old yard management offices demolished
　　　Prefabrication area laid out and concreted
　　　Welding shelters erected on prefabrication area
　　　River dredged in way of no.1 Berth and nos.1 and 2 jetties
　　　Flood bank resurfaced with stone and sprayed with tar 5ft.3in. above whaling
　　　Unit construction commenced on **Stella Orion** (1470)
1962　Complete dredging no.1 and 2 jetties, February
　　　Football club won Barkston Ash cup
　　　Second drawing office extension completed during October
1963　All fitting out of vessels in Hull (Princes Dock) subcontracted to Drypool; no Cochrane employees
　　　　　in Hull
　　　Tower crane erected on bank top in August ; in use from September
　　　Hancosine burning machine installed in August; allowed two identical plates to be burned from a
　　　　　one tenth scale drawing that rotated on a drum under an optical imaging unit
1964　Demolition of fitters' and platers' shops and joiners' store commenced in January
　　　New fitting shop in operation in May
　　　New joiners' shop and sawmill in operation in May
　　　New platers' shop in operation in August
1967　Draughtsman and Allied Technicians Association members locked out March to May
1968　Ross Group opened Selby Fork Hotel, March
1970　Cochrane became part of Drypool Group Shipbuilding Division
1971　Selby yard modernisation continued
1975　Drypool Group in receivership; order book stood at 14 vessels, all completed except 1 tug for Cory,
　　　　　transferred to Richards at Lowestoft
　　　Gales blew down gantry crane, 31 December
1976　Selby yard flooded, 1 January, about 2 feet of water in offices. Some records lost
1977　New 15 ton gantry crane erected over building berths
　　　Staff dinner dance held at York
1981　Closed for royal wedding of Charles and Diana 29 July
1982　Sickness Self Certification for first 7 days introduced
1984　Goole Shipyard acquired from British Shipbuilders
1985　HRH Duchess of Kent launched **Hebridean Isles**
　　　First new building order for Goole Shipyard, **Stirling Esk**
1987　Goole Shipyard closed, labour transferred to Selby
1988　Selective redundancies for 70 yard and staff employees; all on 39 hour week
1989　North British Marine moved to Marina Court Offices
1990　**Danielle of Hamm (157) -** first vessel to be launched by crane
1991　Union Transport cancelled **Union Mercury** and **Union Saturn** through late delivery
　　　Further 70 redundancies
1992　No further support from Howard Smith; last vessel, **Forth Bridge**, delivered end of October
1993　All yard and office equipment sold at auction 25 and 26 March

*A historic moment as the **Forth Bridge** leaves Selby.*

(Cochrane archive)

Biographical details

The following biographical details have a dual purpose. Not only do they provide information about career progression for senior managers in the British shipbuilding industry but they also illustrate how the shipyards comprising that industry were closely interlinked. The details are taken from the *Port of Hull Journal* vol.2, no.2, April 1963, pp.65-7.

Ken Marshall

Ken was born on 14 July 1943 at Cawood, near York. He was educated at the local primary school and Archbishop Holgate School (York). He joined Cochranes, following in the footsteps of his father, and grandfather and commenced work as an apprentice draughtsman in August 1959. He attended Hull College of Technology achieving ONC and HNC results in Naval Architecture, winning the Richard Dunston Design Award for ONC results in 1962 and the Billmeir Shipwrights Award from the Worshipful Company of Shipwrights, for HNC results in 1964. After a spell in the design office, he moved through the estimating and planning departments. He was appointed production manager in 1972 and also was responsible for all subcontracted vessels at Charles Hill in Bristol, Bolson at Poole and Paull shipyard.

He became production director of Cochrane Shipbuilders Limited in January 1978 and was appointed managing director in 1987 until closure of the yard in 1993. He then continued his career in shipping at Richard Dunston (Hessle) Ltd as sales director in 1993/94. He moved to Howard Smith (UK) Ltd. in October 1994 as director of their Felixstowe towage operation. He was appointed to the main board of Howard Smith (UK) Ltd, later Adsteam Ltd, as operations director in 1998 until retirement in September 2005.

Ken was a former member of the executive committee of the Shipbuilders and Repairers Independent Association and Chairman of the inaugural meeting of the European Small Shipyards Group. He was also a former member of the executive committee of the Marine and Allied Industries Training Association. He was a governor of Selby Tertiary College.

A former member of the executive committee of the British Tug Owners Association, he was elected chairman in 2003 and 2004. He had been a member of the executive committee of the European Tug Owners Association and a member of the Germanischer Lloyd British Committee. Having been awarded the Freedom of the City of London, he became a Liveryman of the Worshipful Company of Shipwrights in 2006. Ken Marshall died after a short illness on 28 April 2014, aged 70. He is survived by his wife Kathleen and two children.

Eric Hammal FRINA

Eric joined Wiliam Doxford and Sons (Shipbuilders and Engineers), Sunderland, in 1954 as an apprentice in the ship design office, eventually becoming a design draughtsman. He left Doxfords in 1963 to join Cochrane and Sons Ltd. at Selby as a ship designer and was appointed chief designer in 1966. When Cochrane became part of the Drypool Group, he was promoted to group naval architect being responsible together with the technical director for the development of new vessels of types not previously constructed at the Selby yard. He also prepared designs for construction at the Group's other shipbuilding facilities in Hull and Beverley.

Following the demise of Drypool and when the Selby yard became part of the British Maritime Group he was appointed design manager / principal naval architect for both Selby and Goole. He was responsible for all aspects of design and construction of vessels built within the Group, including conceptual, detail design, preparation of tender documents, liaison with shipowners, attending model tank tests, preparation of specifications and technical calculations with attendance at final performance trials. The vessel types include tankers, dry bulk cargo carriers, reefers, fishing vessels, tugs, offshore support craft, ferries, dredgers etc.

When the shipyard at Selby closed in 1992 he joined North British Maritime Technology at Wallsend-on-Tyne, leaving in 1994 to become principal naval architect for IMT Marine Consultants, setting up a branch of the company in Selby before moving to Escrick near York, employing the experienced personnel from the design/ drawing offices of the old Cochrane yard.

A chartered engineer (C.Eng.), he is still a Fellow of the Royal Institute of Naval Architects and was formerly a member of Lloyds Register of Shipping Technical Committee. Now retired, he lives in Sheffield, producing watercolour paintings of ships.

An important source book written from the viewpoint of a naval architect is : *Cochranes of Selby - Yorkshire Shipbuilders 1884-1992*, written by Eric Hammal and Peter D Coates, 278pp, and published by York Publishing Services in 2013.

Douglas M Waite, managing director and shipyard director

Mr Waite joined Cochrane as general manager in 1961. He had been apprenticed to Fairfield Shipbuilding and Engineering Co and became draughtsman with Charles Connell and Sons,

Glasgow, and the Caledon Shipbuilding and Engineering Co, Dundee. He then joined Alexander Stephen and Son and became repair manager. After a period as shipbuilding manager with William Hamilton and Co, Port Glasgow, he went to Canada as production manager for the Saint John Shipbuilding and Dry Dock Co, New Brunswick, and left them to join Cochrane.

A Clifford Kelly, technical director

Mr Kelly had been an indentured apprentice as a shipbuilders draughtsman with John I Thornycroft and Co Ltd, Southampton. He later joined J Samuel White, of Cowes, and Vospers, of Portsmouth. In the war years, he was engaged in the construction of coastal patrol vessels. He went to Cochrane in 1951 as naval architect and was appointed technical director in 1960.

W M George, B.Sc, naval architect and technical director.

Mr George was aged 35 when he arrived from

Tyneside, and had served a student apprenticeship at the Neptune Yard of Swan Hunter and had studied at King's College, Newcastle. He became design section leader at William Doxford and Sons, Sunderland. He joined Cochrane from Burness, Corlett and Partners Ltd, Basingstoke.

Robert J Coles, shipyard manager

Aged 42 and a native of Somerset, he was an indentured apprentice with Charles Hill and Sons, Ltd, Bristol, from 1935 until 1941. In 1941 he became ship manager, and in 1945 was promoted to assistant shipyard manager. He joined Cochrane in October 1948 as assistant shipyard manager and was promoted to shipyard manager in 1960.

R C Nelson, company secretary.

Aged 28, he had been an apprentice draughtsman at Blackburn Aircraft, Brough, between 1949 and 1951. In 1953 he joined Hodgson, Harris & Co, accountants. He was loaned to Cochrane in 1959 and appointed secretary in 1962.

A presentation in the drawing office. Sadly, neither the surname of the recipient nor the reason for the presentation are known. From left to right: A L Cochrane, Ken Marshall, Bryn Walker (partly hidden), Cliff Kelly, Paul Hendrie, Mike ??, Harry Taylor, Fred McGuire, Bill George, Reg Walker.

(Cochrane archive)

The Cochrane Archive

After the closure of the Cochrane shipyard in 1992, Arthur Credland contacted Ken Marshall, the managing director, to ask about the future of the company archives. He was very keen that the collection should be kept together and he was pleased that Hull Maritime Museum was willing to find space for this record of a major local shipbuilding company. A large van paid three visits to Selby to collect the material, the largest bulk comprising thousands of plan rolls, hand drawn on linen. The photographs, plans, ledgers and cost estimate books

were sorted. It was then realised that it would be possible to compile an illustrated yard history. Arthur Credland retired from the Hull museum in 2008 and the entire archive was transferred to the North Yorkshire County Record Office in Northallerton as Selby is in the administrative area of North Yorkshire. A strenuous effort was then made to find a publisher and so the saga which began more than twenty years ago has culminated in three volumes, published between 2012 and 2016.

ALEXANDRA
TOWING CO. LTD.
LIVERPOOL

BOYD LINE LTD.
HULL

BUNCH STEAM
FISHING CO. LTD.
GRIMSBY

CONSOLIDATED
FISHERIES LTD.
GRIMSBY

CONSOLIDATED
FISHERIES LTD.
GRIMSBY

GASELEE & SON LTD.
LONDON

LORD LINE LTD.
HULL

J. MARR & SON LTD.
FLEETWOOD

MILFORD STEAM
TRAWLING CO. LTD
MILFORD HAVEN

NORTHERN
TRAWLERS LTD
GRIMSBY

RINOVIA STEAM
FISHING CO. LTD.
GRIMSBY

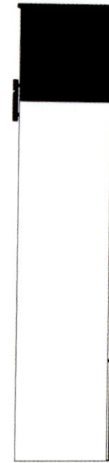

ST. ANDREWS STEAM
FISHING CO.LTD.
HULL

TILBURY CONTRACTING
& DREDGING CO. LTD.
LONDON

UNITED
TOWING CO. LTD.
HULL

WEST DOCK STEAM
FISHING CO. LTD
HULL

BOWKER & KING
LONDON

BRITISH UNITED
TRAWLERS
GRIMSBY

CALEDONIAN
MACBRAYNE LTD.
GLASGOW

CLARIDGE
TRAWLERS LTD.
LOWESTOFT

CONSOLIDATED
FISHERIES LTD.
GRIMSBY

COUNTY SHIPS LTD.
LONDON

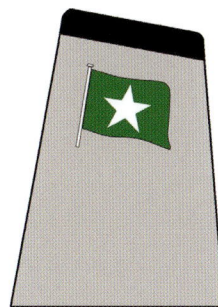

DERWENT
TRAWLERS LTD.
GRIMSBY

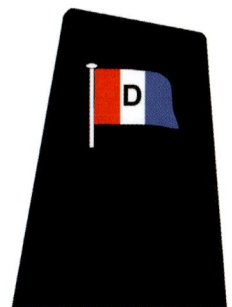

DOMINION STEAM
FISHING CO. LTD.
GRIMSBY

ELLIOT STEAM
TUG CO. LTD.
LONDON

ESSO PETROLEUM
CO. LTD.
LONDON

FORTH
TANKERS PLC.
EDINBURGH

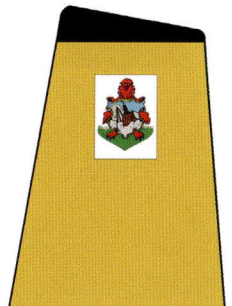

GOVERNMENT OF BERMUDA
MARINE & PORTS
HAMILTON BERMUDA

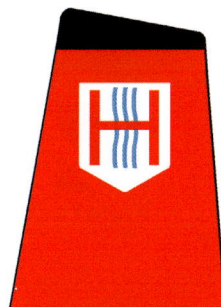

HARRISON'S
(CLYDE) LTD.
GLASGOW

JAPAN
FISHING CO. LTD.
GRIMSBY

JÖRUNDUR H/F
REYKJAVIK
ICELAND

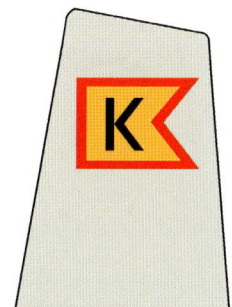

KLONDYKE
SHIPPING CO. LTD.
HULL

LONDON & ROCHESTER
TRADING CO. LTD.
LONDON

METCALF MOTOR
COASTERS LTD.
LONDON

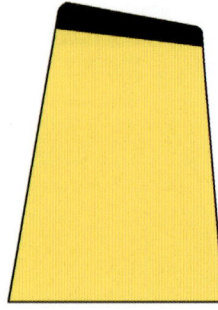

MINISTRY OF AGRICULTURE
FISHERIES & FOOD
LOWESTOFT

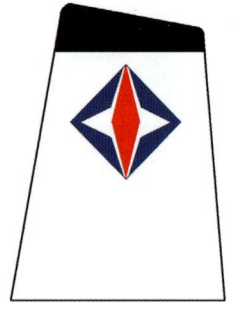

NORTH BRITISH
SHIPPING LTD.
HULL

ORKNEY FERRIES LTD.
KIRKWALL

REA TOWING CO. LTD.
LIVERPOOL

ROWBOTHAM & SONS
(MANAGEMENT) LTD.
LONDON

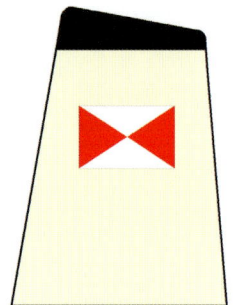

SCOTTISH
NAVIGATION CO. LTD
LONDON

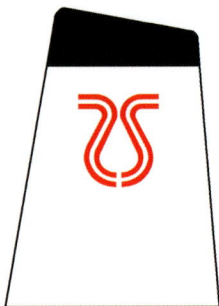

SEAFORTH
MARITIME LTD.
ABERDEEN

SEALINK ISLE OF
WIGHT FERRIES
LONDON

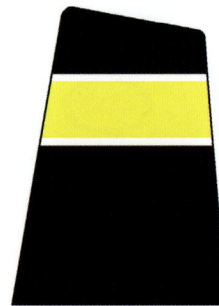

SHELL MEX & BP LTD
LONDON

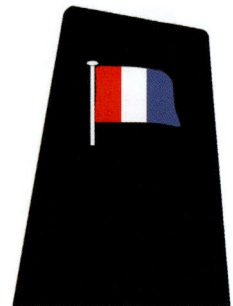

SIR T. ROBINSON & SON
(GRIMSBY) LTD.
GRIMSBY

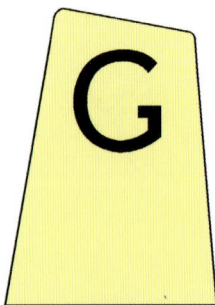

TALISMAN
TRAWLERS LTD.
WEST HARTLEPOOL

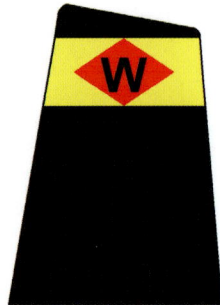

J. WHARTON
SHIPPING LTD.
GOOLE

WYRE TRAWLERS LTD.
FLEETWOOD

ZAPATA OFFSHORE
SERVICES LTD.
GREAT YARMOUTH

Abbreviations and Explanations

Armament

1-12pdr (6pdr, 3pdr) - A single mounting usually a Hotchkiss delivering a shell of the poundage nominated. In some vessels in the latter period of WW2, the quick firing - QF 3pdr Hotchkiss or the Ordinance QF 3pdr Vickers gun was fitted.

1-4" - A single mounting usually a quick firing - QF 4-inch MkXIX.

Number-20mm AA; Number-0.5" AA - Mountings capable of engaging aircraft.

HA - High Angle. A mounting enabling the gun to engage aerial as well as surface targets.

AA - Anti Aircraft - A mounting capable of engaging aircraft.

A/S - Ship fitted out for anti-submarine warfare.

M/S - Ship fitted out for minesweeping duties.

1-7.5" A/S Howitzer - A single mounting often referred to as a 'bomb thrower' delivering a projectile to explode in the water in pursuit and disabling of submarines.

W/T - Equipped with wireless telegraphy receiver/transmitter.

DC - Equipped with Depth Charges to engage submarines, delivered either by rails and ramp or thrower.

Pennant Numbers (Allocated to Cochrane built vessels)

P.No. FY - The FY pennant followed by the allotted number. Allocated requisitioned trawlers engaged in A/S - Anti submarine; A/P -Auxiliary patrol; M/S - Minesweeping; D/L - Dan laying work and some trawlers/drifters employed on the balloon barrage.

P.No. T - T flag superior followed by the allotted number. Allocated to Royal Navy A/S and M/S trawlers.

P.No. W - W flag superior followed by the allotted number. Allocated to tugs and salvage vessels.

P.No. Y - Y flag superior followed by the allotted number. Allocated to stores, fuel and water carriers.

P.No. Z - Z flag superior followed by the allotted number. Allocated to some, but not all trawlers and drifters employed on boom defence, gate and net carrying/laying duties.

P.No. 4 - Number 4 flag superior followed by the allotted number. Allocated to auxiliary patrol trawlers, some balloon barrage, torpedo recovery, wreck detection/dispersal, examination service and armed boarding vessels.

Machinery, Propulsion and Fit-out

AZP - Fixed pitch propeller mounted in a steerable pod.
CP - Controllable pitch propeller.
NHP - Nominal horsepower calculated by formula.
T.3-cyl - Triple cylinder direct acting compound vertical steam engine.
VSP - Voith Schneider cycloidal propeller, combining propulsion and steering in one unit.

BHP - Brake horsepower calculated by application.
IHP - Indicated horsepower calculated by application.

Service

RASC - Royal Army Service Corps
RNR - Royal Naval Reserve

RN - Royal Navy
RNVR - Royal Naval Volunteer Reserve.

Ranks

Sk - Skipper
Lt - Lieutenant

Ty - Temporary
Lt Cdr - Lieutenant Commander

Act - Acting
Cdr - Commander

Miscellaneous

BoT - Board of Trade, British Government Ministry.

C&M - A vessel laid up but hull and machinery maintained to enable rapid re-activation.

IMO - International Maritime Organisation.

M.O.T. - Ministry of Transport, British Government Ministry.

M.O.W.T. - Ministry of War Transport, British Government Ministry during WW2.

RSS - Registrar of Ships and Seamen, British Government Department.

U.N.R.&R. Scheme - United Nations Relief & Rehabilitation Scheme. Set up in 1943 and administered from 1945 by the United Nations to provide relief in the form of food and medicine to areas liberated from Axis powers after WW2 ended.

WFA - White Fish Authority, part of UK Ministry of Agriculture & Fisheries.

Trophies

Prunier Trophy (Prunier Herring Trophy) : Made from Purbeck marble and offered for an annual competition for fishing boats (drifters) by London restaurant owner Madame Prunier in 1936 to promote the eating of herring. Awarded to the boat which had caught the largest haul of herring with one shot, in one night between dates set by a committee and landed at Gt Yarmouth or Lowestoft.

The Distant Water Challenge Shield : The Hull Distant Water Challenge Shield was awarded from 1967 to 1977 to the skipper and crew of the trawler with the greatest value of catch for the year, on a points basis. Points calculated by adding kits landed to gross returns in £s and dividing by the registered speed of the vessel.

The Silver Cod Challenge Trophy : Awarded from 1954 to 1968 by the British Trawler Federation to the skipper and crew of the vessel with the largest total landed catch for the year.

NAME Official no. PLN vessel type	Yard no. Ordered Laid down	Launched Completed	Registered G ton N ton	Length ft Beam ft	Engine builder Horse power Registered speed	OWNER
ASH T39 Tree class	1208 05.06.1939 31.07.1939	13.11.1939 06.05.1940	550 disp	150.0 27.5 10.5	Amos & Smith 850ihp 3-cyl 11.5 knots	The Admiralty, Whitehall, London

Commissioned as a minesweeper (1-12pdr, AA weapons) (P.No.T39). Based Portsmouth with 25th M/S Group. 05.06.1941: Sweeping in Thames estuary (Lt Arthur G Newell, RNZNVR) in company with HMS BIRCH (P.No.T93). Mined ½ mile SW of Knock John Buoy; survivors and many fittings taken onboard BIRCH and with HM Tug MAMMOTH (P.No.W56) alongside attempted to reach Sheerness. Pumps failed, started to settle and foundered in position 51.30N 01E. Two crew killed and six injured.

| **BAY**
T77
Tree class | 1209
05.06.1939
15.08.1939 | 12.12.1939
10.06.1940 | 550 disp | 150.0
27.5
10.5 | Amos & Smith
850ihp 3-cyl
11.5 knots | The Admiralty,
Whitehall,
London |

Commissioned as a minesweeper (1-12pdr, AA weapons) (P.No.T77). Based Portsmouth with 25th M/S Group. 1947: Sold to Norwegian interests. 1951: Converted to motor and fitted with 6-cylinder oil engine by Wumag-Krupp, Waggon & Maschinenbau, Hamburg. 1952: Sold to Tristan da Cunha Development Co Ltd, Tristan da Cunha. Renamed ISOLDA but not registered. 1952: Registered at Cape Town as **TRISTANIA**. Official No.191382, 628grt, 345net. 10.10.1961: Along with the wooden mfv FRANCES REPETTO (316grt/1944) played an important roll in the total evacuation of 264 islanders and 26 expatriates to Nightingale Island from Tristan da Cunha during the volcanic eruption. 1963: Sold to Tristan Investments (Pty) Ltd, Cape Town. Converted to fish factory vessel, re-engined with 12-cylinder 900bhp Brons oil engine by Drypool Eng & Dry Dock Co Ltd, Hull. 1973: Sold to Marine & Industrial Cleaning, Cape Town. 1976: Sold to Oceanpac (Pty) Ltd, Walvis Bay, SW Africa. 1977: Sold to Kuttlefish S W A (Pty) Ltd, Walvis Bay. 1977: Laid up in Cape Town following a fire. 22.02.1978: Scuttled off Robben Island, Cape Town, for use as a recreational dive site. 1978: Registry closed.

| **IWATE**
167797
CF5   Steam trawler | 1210
1939
1939 | 30.09.1939
11.01.1940 | 314
116 | 130.7
24.5
11.9 | Amos & Smith
125nhp 3-cyl
12.2 knots | Neale & West Ltd,
Cardiff |

01.1940: Registered at Cardiff (CF5). Fitted with AA weapons and complement increased by two DEMS gunners. 06.09.1940: Fishing off Co Galway coast in position 53.30N 14.00W, suffered damage from German air attack; no casualties. 03.10.1940: Outward to fishing grounds when some 5 miles NW of Mizen Head, Co Cork, strafed and bombed by German Kondor aircraft with some damage; mate and deckhand injured. Put into Castletownbere for temporary repairs and treatment of wounded. 11.11.1940: When some 35 miles SW of Old Head of Kinsale, Co Cork, bombed and strafed by German aircraft; some damage but no casualties. 23.12.1940: On SW of Ireland grounds, bombed by German aircraft in approximate position 52.55N 12.30W; damaged but no casualties. 18.09.1944: Sailed Cardiff for Porcupine Bank (Sk Rhymer) to determine quantity and quality of fish after limited exploitation in war years. 02.10.1944: At Cardiff landed 1125cwt of fish including 524cwt of hake and 355cwt of haddock. 01.05.1946: Outward from Cardiff for the fishing grounds when some 25 miles south-west of Old Head of Kinsale, struck a mine. All crew took to boat and vessel foundered. All survivors picked up. Registry closed.

| **LE ROYAL**
166638
GY400   Motor trawler | 1211
1940
1940 | 26.07.1940
12.1941 | 316
94 | 142.6
24.6
13.6 | Ruston & Hornsby
750bhp 8-cyl
12.0 knots | Grimsby Motor Trawlers Ltd,
Grimsby |

First motor trawler designed for Arctic fishing. Completed at a cost of £28,500.
01.10.1940: Requisitioned from builder but completion delayed by late supply of main engine. 12.1941: Completed as a minesweeper (1-12pdr, AA weapons, DC); hire rate £240.19s.0d per month. 31.12.1941: Registered at Grimsby (GY400). 01.1942: Renamed POSTBOY (P.No.FY1750). Mediterranean station. 10.02.1944: Sold to Milford Steam Trawling Co Ltd, Milford Haven. 27.12.1944: Grimsby registry closed. 30.12.1944: Registered at Milford as **LE ROYAL** (M14). 07.1946: Returned. Registered at Milford as **MILFORD MARQUIS** (M14). 06.1951: Sold to N V Vissch Onderneming 'de Vem', IJmuiden, Holland. 13.06.1951: Milford registry closed. 08.1951: Registered at IJmuiden as **POST BOY** (IJM35). 21.08.1951: Unslipped after survey and overhaul. 1964: Sold to Claridge Trawlers Ltd, Lowestoft. IJmuiden registry closed. 27.01.1964: Registered at Lowestoft. 28.01.1964: Registered at Lowestoft as **ST KITTS** (LT481). 1965: Lowestoft top trawler (Sk Ernest Peek) - £61,209. 04.10.1974: Towed into Lowestoft after sustaining major engine damage and laid up. 09.1976: Sold for breaking up. 28.09.1976: Sailed Lowestoft for Humber in tow of standby motor vessel ST MARTIN (254grt/1961) also for breaking up. 11.1976: Broken up on the Humber. 03.11.1977: Lowestoft registry closed except in respect of mortgage. 30.03.1979: Registry closed on discharge of mortgage.

| **Not built** | 1212   Order cancelled | | | | | |

| **TRITON**
168023
Lighthouse tender | 1213
1939
1939 | 27.02.1940
09.08.1940 | 680
232 | 173.2
29.1
14.6 | Amos & Smith
83nhp 3-cyl
11.7 knots | Hon Corporation of the Trinity House,
London |

Ordered by Boston Deep Sea Fishing & Ice Co Ltd, Fleetwood, as a steam trawler. Launched as **QUEEN OF THE WAVES**, 575grt. Sold to the Hon Corporation of the Trinity House, London, and completed as a lighthouse tender; 680grt, 232net. Registered at London as **TRITON**. 1940: Sold to the Ministry of War Transport and fitted out as a light vessel relief tender. 1945: Sold to the Hon Corporation of the Trinity House, London. 1957: Took the part of the Chinese ship KIANLANG LIBERATOR in the filming of the "Yangtze Incident". 1963: Sold to Belgian shipbreakers. 24.07.1963: Sailed Harwich for Bruges in tow of Belgian motor tug MARTINE LETZER (161grt/1960). 07.1963: Registry closed.

Bay (1209), a Tree class minesweeper.

(Cochrane archive)

The lighthouse tender **Triton** *(1213).*

(Cochrane archive)

FANDANGO T107 Dance class	1214 09.09.1939 24.10.1939	26.03.1940 11.07.1940	530disp	150.0 27.6 10.5	Amos & Smith 850ihp 3-cyl 11.5 knots	The Admiralty, Whitehall, London

Commissioned as an anti submarine trawler/escort (1-4", AA weapons, ASDIC, DC) (P.No.T107). 21.09.1940: Escorting convoy HX-72 (Ty/Lt F C Hopkins RNVR). Picked up fifteen survivors from Glasgow registered tanker INVERSHANNON (9154grt/1938) (Capt William R Forsyth), torpedoed by U-boat (U99) 480 miles W of Bloody Foreland in position 50.40N 22.04W. 29.09.1940: Landed survivors at Belfast. 30.05.1943: Escort duty on East African coast (Ty/Lt A G Fisher RNVR). Escorting American 'Liberty' steamer FLORA MACDONALD (7177grt/1943) (Capt Ernest W Jones) Liberia to Freetown, when off Freetown torpedoed by U-boat (U126) and set on fire. Sixty-three survivors picked up and landed Freetown. FLORA MACDONALD taken in tow by HM Tug ZWARTE ZEE (P.No.W163) towed to Freetown and beached, some cargo salvaged but vessel burnt out. Total loss. 1946: Sold to Skibs A/S Argo Tvedestrand, Norway. Registered at Tvedestrand as **FANDANGO**. Remeasured to 452grt, 142net. 1948: Converted to refrigerated cargo motor vessel and fitted with 8-cylinder oil engine by Crossley Bros Ltd, Manchester. Remeasured to 493grt, 223net. 1957: Re-engined with 6-cylinder 750bhp oil engine by Klöckner-Humboldt-Deutz, Cologne - 10½ knots. 1966: Sold to A/S Samfrost, Stavanger. Tvedestrand registry closed. Registered at Stavanger as **SAMFROST**. 23.07.1972: Vessel extensively damaged by fire at Stavanger. Sold to Brodrene Anda for breaking up with refrigerated machinery removed for fitting in replacement vessel. 1972: Registry closed.

FOXTROT T109 Dance class	1215 09.09.1939 24.10.1939	23.04.1940 29.08.1940	530disp	150.0 27.6 10.5	Amos & Smith 850ihp 3-cyl 11.5 knots	The Admiralty, Whitehall, London

Commissioned as an anti submarine trawler/escort (1-4", AA weapons, ASDIC, DC) (P.No.T109). 1941: Based Belfast with A/S Convoy Escort Group. 1943: Transferred to the Mediterranean Station on North African convoy duties. 20.04.1943: Off North African coast (Ty/Lt Juan B Bald RNVR). With HMS STELLA CARINA (P.No.FY352) (Lt John Vernon Lobb RANVR) (see yard no.1165) assisted in the rescue of 61 crewmen (who had been in the water for two hours) from the American merchant ship MICHIGAN (5594grt/1919) torpedoed by U-boat (U565) in convoy UGS-7 off Oran, Algeria. 31.07.1946: Transferred to the War Dept. 1947 - 1950: Based Birkenhead with Royal Army Service Corps; employed as escort for vessels dumping ammunition. 1951: Sold to BISCO and allocated to Thos W Ward Ltd, Sheffield, for breaking up. 30.09.1951: Arrived Barrow in tow.

ASSURANCE W59 Assurance class tug	1216 16.10.1939 18.12.1939	23.05.1940 19.09.1940	597disp 5	142.6 33.3 11.0	C D Holmes 1350ihp 3-cyl 13.0 knots	The Admiralty, Whitehall, London

Commissioned (1-3", AA weapons) (P.No.W59). 20.05.1941: In rescue tug role (Sub Lt E E Litts RN) with HM Ships FAULKNER (P.No.H62) (Capt A F de Salis, RN) and LINCOLN (P.No.G42) (Lt R J Hanson RN), picked up fifty eight survivors from the British motor ship JAVANESE PRINCE (8583grt/1925) (Capt George Gillanders) which had been sunk by torpedoes from U-boat (U138) 155 miles NW of the Butt of Lewis, Outer Hebrides. Survivors transferred to rescue ship TOWARD (1571grt/1923) (Capt A J Knell). 28.05.1941: Survivors landed at Gourock. 18.10.1941: Stranded in Lough Foyle, Northern Ireland; extensive bottom damage. Declared a total loss.

DILIGENT W18 Assurance class tug	1217 16.10.1939 27.12.1939	22.06.1940 15.10.1940	597disp 5	142.6 33.3 11.0	C D Holmes 1350ihp 3-cyl 13.0 knots	The Admiralty, Whitehall, London

Commissioned (1-3", AA weapons) (P.No.W18). 1940: Renamed **TENACITY** (W18). 16.01.1941: In rescue tug role with HMS WESTCOTT (P.No.D47) (Lt Cdr W F R Segrave RN) and HM Tug SUPERMAN (P.No.W89), picked up 143 survivors from the British passenger ship OREPESA (14118grt/1920) which was torpedoed at 3.56am by U-boat (U96) SE of Rockall in position 52 28N 12W. Two further torpedo hits caused her to capsize and sink at 5.59am. 1947: Based Sheerness Dockyard. Renamed **ADHERENT**. 1962: Sold to Bergnings & Dykeri A/B Neptun, Stockholm, Sweden. Registered at Stockholm as **HERMES**. 1962: Converted to motor and fitted with two 9-cylinder oil engines by Ruston & Hornsby, Lincoln, geared to single shaft. Re-classed as motor salvage tug, 567grt - 10knots. 1970: Sold to Rivtow Marine Ltd, Vancouver, British Columbia, Canada. Stockholm registry closed. Registered at Vancouver as **RIVTOW VIKING**. Official No. 330812. 1973: Company restyled Rivtow Straits Ltd, Vancouver. 1982: Sold to Rivtow Industries Ltd, Vancouver. 03.1983: Laid up. 1986: Sold to Bill Church, Vancouver. Registered at Vancouver as **CANADIAN VIKING** (IMO 5149136). 1987: Sold to Viking Overseas Leasing Ltd, Richmond, BC. 06.2015: Still in service.

PRUDENT W73/A85 Assurance class tug	1218 16.10.1939 11.03.1940	06.08.1940 19.11.1940	597disp 5	142.6 33.3 11.0	C D Holmes 1350ihp 3-cyl 13.0 knots	The Admiralty, Whitehall, London

Commissioned (1-3", AA weapons) (P.No.W73). Based Iceland and later Shetland. 1947: Registered at Harwich as **CAUTIOUS**. Official No.182199. Manned by Royal Fleet Auxiliary (P.No.A385). 09.1964: For sale at Chatham. 12.1964: Sold to M R Cliff Tugboat Co Ltd, Vancouver, British Columbia, Canada. Harwich registry closed. Registered at Vancouver as **RIVTOW LION**. 1965: Converted to motor and fitted with 8-cylinder 3200bhp oil engine by Stork NV, Amsterdam. 1973: Sold to Rivtow Straits Ltd, Vancouver. 1985: Towed to Mosquito Creek, North Vancouver, and used as a breakwater to protect the marina. Later leaking oil and towed to Ladysmith Harbour, Vancouver Island, for tank cleaning. 2002: Sold to Nanaimo Dive Association (NDA), Nanaimo, and prepared for scuttling. 2003: Registry closed. 2003: Sold to David Charles Cobb, New Westminster, BC. 06.02.2005: Scuttled in Departure Bay, Nanaimo, Vancouver Island, BC as a shallow water artificial reef and recreational dive site.

RESTIVE W39/A286 Assurance class tug	1219 16.10.1939 25.04.1940	04.09.1940 10.12.1940	700disp	142.6 33.3 11.0	C D Holmes 1350ihp 3-cyl 13.0 knots	The Admiralty, Whitehall, London

Commissioned (1-3", AA weapons) (P.No.W39). Based Mediterranean (Lt D M Richards RNR). 01.12.1941: Took over tow of HMS MANXMAN (P.No.M70) torpedoed and badly damaged by U-boat (U375) off Algiers when on passage Algiers towards Gibraltar. 02.12.1941: Delivered Mers el-Kébir, Oran, for emergency repairs. 21.12.1942: Took over tow of troopship STRATHALLAN (23,722grt/1938) which had been torpedoed by U-boat (U562) at 2.23am when convoy KMF-5 was some 40 miles north of Oran. Proceeded towards Oran but explosion and fire ensued onboard and remaining crew taken off. After further fourteen hours under tow. vessel started to settle and foundered 12 miles off Oran in position 36.01N 00.33W. 24.02.1943: At 9.00pm took over tow of American 'Liberty' ship NATHANAEL GREEN (7176grt/1942) torpedoed by U-boat (U656) when convoy MKS-8 was some 40 miles northwest of Oran. Later further damaged by air attack with aerial torpedo. 25.02.1943: Successfully beached vessel at Salamanda 4 miles west of Mostaganem; total loss but cargo salved. 04.07.1943: With other vessels picked up crew and troops from British ships CITY OF VENICE (8762grt/1924) and ST ESSYLT (5634grt/1941) which had been torpedoed by U-boat (U357) at 9.40pm when convoy KMS-18B was off Cape Tenez, Algeria. 08.04.1947: Transferred to Bermuda. 28.07.1948: Paid off at Portland. Based Portsmouth with CD. 11.10.1954: Based Portland target towing (P.No.A286). 18.11.1964: Arrived Pembroke Dock, destored and laid up pending sale. 11.06.1965: Sold to Branco Salvage Ltd, Famagusta, Cyprus. Registered at Famagusta as VENTURA. Official No.321016. 597grt 5net. 1966: Sold to Goldsworthy Mining Pty Ltd, Fremantle, Australia. Famagusta registry closed. Registered at Fremantle as NULLAGINE. 1971: Sold to Brunei Shipping & Shipbuilding Panama S.A., State of Brunei. Fremantle registry closed. Registered at Panama as MAN SOON. 1974: Sold to Top Service Inc, Manila, Philippines. 1999: Deleted from Lloyd's Register of Shipping - "Vessel's continued existence in doubt".

CELIA T134 Shakespearean class	1220 12.12.1939 25.05.1940	18.09.1940 18.01.1941	545disp	150.0 27.9 11.0	Amos & Smith 950ihp 3-cyl 12.0 knots	The Admiralty, Whitehall, London

Commissioned as a minesweeper (1-12pdr, AA weapons) (P.No.T134). Home Fleet with M/S Group 73. 1946: Sold to Armand Vella, Marseilles, France. Converted to cargo motor vessel and fitted with 8-cylinder 800bhp oil engine by Crossley Bros Ltd, Manchester. 489grt 221net. Registered at Marseilles as CELIA. 1958: Sold to Jacques Prat, Marseilles. Registered at Marseilles as VANDA. 1961: Sold to Ong Kin Hock, Penang, Malaya. Marseilles registry closed. Registered at Penang as KUALA KANGSAR. Official No.199549. Remeasured to 439grt, 237net, 690dwt. 1966: Sold to Hup Hoe Shipping Co, Penang. 19.03.1969: Foundered in the outer harbour at Singapore on loaded voyage towards Penang. One crew lost. 04.1969: Registry closed.

CORIOLANUS T140 Shakespearean class	1221 12.12.1939 04.05.1940	02.10.1940 05.02.1941	545disp	150.0 27.9 11.0	Amos & Smith 950ihp 3-cyl 12.0 knots	The Admiralty, Whitehall, London

Commissioned as a minesweeper (1-12pdr, AA weapons) (P.No.T140). Based Falmouth with M/S Group 74. 11.1942: Took part in Allied North Africa landings. 07/08.1943: Took part in Operation Husky - Sicily landings. 09.1943: Took part in Operation Avalanche - Salerno landings. 01.1945: Employed in northern Adriatic. 05.05.1945: On patrol in Northern Adriatic (Ty/Lt N Hunt RNVR) mined and foundered six miles from Novigrad/Cittanova in position 45.19N 13.25E. All crew saved. 07.05.1945: Unconditional surrender of all German forces to the Allies.

FLUELLEN T157 Shakespearean class	1222 12.12.1939 09.07.1940	01.11.1940 18.03.1941	545disp	150.0 27.9 11.0	Amos & Smith 950ihp 3-cyl 12.0 knots	The Admiralty, Whitehall, London

Commissioned as a minesweeper (1-12pdr, AA weapons) (P.No.T157). Based Falmouth with M/S Group 74. 1942: Mediterranean Station. 09.11.1942: Off Oran, Algeria (Ty/Lt Basil J Hampson RNR) in collision with HMS GARDENIA (P.No.K99) which foundered in position 35.49N 01.05W; three crew lost. 12.11.1942: Escorting convoy KMS-2 (Ty/Lt John H Hill RNR). Picked up sixty-one survivors from the Liverpool steamer BROWNING (5332grt/1919) (Capt Thomas J Sweeney) torpedoed and set on fire by U-boat (U593). After abandonment, further explosions and vessel foundered in position 35.53N 00.33W; one crew lost. Survivors landed at Oran. 1947: Transferred to Scottish Home Department, Edinburgh. 1948: Renamed SCOTIA (unregistered), 492grt, 138net. Employed as a research vessel. 1965: Transferred to Department of Agriculture & Fisheries for Scotland. Registered at Leith as SCOTIA. Official No.303097. 01.1972: Registered at Leith as SCARBA freeing name for new build. Fishery protection duties. 05.1973: Laid up. 11.1973: Sold to W H Arnott Young & Co (Shipbreakers) Ltd, Dalmuir, for breaking up. 1974: Registry closed.

CANNA T161 Isles class trawler	1223 06.04.1940 19.07.1940	18.11.1940 04.04.1941	545disp	150.0 27.6 10.5	C D Holmes 850ihp 3-cyl 12.0 knots	The Admiralty, Whitehall, London

Commissioned as a minesweeper (1-12pdr, AA weapons) (P.No.T161). Based Lagos with 1st Trawler Group. 05.12.1942: British tanker ATHELVICTOR (8410grt/1941) discharging petroleum spirit in Lagos Harbour suffered spillage which ignited causing a large explosion resulting in the loss of HM Ships CANNA (Lt W N Bishop-Leggett RNR), twelve crew lost; BENGALI (P.No.FY165) (Lt R S Penby RNR) 455grt/1937, one crew lost, and SPANIARD (P.No.FY144) (Lt J B Love RNR) 455grt/1937, two crew lost.

COPINSAY T147 Isles class trawler	1224 27.05.1940 07.08.1940	02.12,1940 24.04.1941	545disp	150.0 27.6 10.5	Amos & Smith 850ihp 3-cyl 12.0 knots	The Admiralty, Whitehall, London

Commissioned as a minesweeper (1-12pdr, AA weapons) (P.No.T147). Based Lagos with 1st Trawler Group. Minesweeping between Freetown and Lagos. 04.1946: Transferred to War Department, London; RASC fleet. Fitted out as a cargo vessel employed dumping ammunition, etc. 1956: Sold to E Abbot & Ph Georiadis, Piræus. Registered at Piræus as ION (445grt, 240net). 31.12.1958: In collision north of Crete and subsequently foundered. 1959: Registry closed.

CUMBRAE	1225		545disp	150.0	Amos & Smith	The Admiralty,
T154	22.07.1940	30.12.1940		27.6	850ihp 3-cyl	Whitehall,
Isles class trawler	02.09.1940	22.05.1941		10.5	12.0 knots	London

Commissioned as a minesweeper (1-12pdr, AA weapons) (P.No.T154). Mediterranean Fleet with 2nd Trawler Group. 1942: Based Haifa as convoy escort. 22.01.1946: Sold to Italian Navy (Marina Militare) (P.No.RD302). 1965: Expended as a target.

FARA	1226		545disp	150.0	Amos & Smith	The Admiralty,
T162	27.05.1940	15.01.1941		27.6	850ihp 3-cyl	Whitehall,
Isles class trawler	05.09.1940	24.06.1941		10.5	12.0 knots	London

Commissioned as a minesweeper (1-12pdr, AA weapons, DC) (P.No.T162). Based Mombasa with 3rd Trawler Group. 20.02.1944: Off the Maldives (Ty/Lt J M Walton RNVR) with HM Trawler OVERDALE WYKE (P.No.FY338) (see yard no.937), under direction of an RAF Catalina flying boat, picked up 36 survivors from Greek steamer EPAMINONDAS C EMBIRICOS (4385grt/1927) (Capt Constantinos A Marmaras) torpedoed by U-boat (U168) on 15.02.1944 north of Addu Atoll, Maldives. 21.02.1944: Landed survivors at Addu Atoll. 12.07.1946: Sold and renamed **DIKSMUIDE 7**. No further details.

FLOTTA	1227		545disp	150.0	Amos & Smith	The Admiralty,
T171	27.05.1940	13.02.1941		27.6	850ihp 3-cyl	Whitehall,
Isles class trawler	04.10.1940	05.06.1941		10.5	12.0 knots	London

Commissioned as a minesweeper (1-12pdr, AA weapons, DC) (P.No.T171). 29.10.1941: Stranded on Buchan Ness (Ty/Sub Lt A Smith RNVR). Skeleton crew remained onboard with a view to refloating. 06.11.1941: Came afloat but making water, drifted and foundered 3 miles ESE of Buchan Ness lighthouse. Peterhead lifeboat JULIE PARK BARRY OF GLASGOW (Cox John B McLean) attended and took off nine crew; one crew lost. Wreck lies in 62m in position 57.27N 01.41W.

RONALDSAY	1228		545disp	150.0	Amos & Smith	The Admiralty,
T149	22.07.1940	15.02.1941		27.6	850ihp 3-cyl	Whitehall,
Isles class trawler	07.10.1940	17.07.1941		10.5	12.0 knots	London

Commissioned as a minesweeper (1-12pdr, AA weapons, DC) (P.No.T149). Based Scapa Flow with 4th Trawler Group. 27.08.1946: Sold to Tin Tai Navigation Co, Shanghai, China. 1948: Converted to dry cargo vessel. Registered at Shanghai as **DAH LAI**. 486grt, 175net. 1992: Removed from Lloyd's Register of Shipping - "Vessel's continued existence in doubt".

RYSA	1229		545disp	150.0	Amos & Smith	The Admiralty,
T164	27.07.1940	15.03.1941		27.6	850ihp 3-cyl	Whitehall,
Isles class trawler	19.11.1940	13.08.1941		10.5	12.0 knots	London

Commissioned as a minesweeper (1-12pdr, AA weapons, DC) (P.No.T164). Based Portland with 5th Trawler Group. 10-12.1943: Italian campaign. 12.1943: Following surrender of Italian fleet, on escort duties (Ty/Act/Lt Cdr John Handley Cooper RNVR). 08.12.1943: Mined off La Maddalena Island, Sardinia, and foundered with the loss of the CO and eighteen crew.

SHAPINSAY	1230		545disp	150.0	Amos & Smith	The Admiralty,
T176	13.09.1940	29.03.1941		27.6	850ihp 3-cyl	Whitehall,
Isles class trawler	19.11.1940	04.09.1941		10.5	12.0 knots	London

Commissioned as a minesweeper (1-12pdr, AA weapons, DC) (P.No.T176). Based Plymouth with 6th Trawler Group. 1942: Based Mombassa 10.08.1946: Sold to Scerif Abo' Imanchio, Mogadiscio, Italian Somaliland. Registered at Mogadiscio as **EL HASCIMY**. 1949: Converted to dry cargo; lengthened to 163.4ft; remeasured to 480grt, 180net, and converted to burn oil fuel. 1955: Sold to Hussein Fayez, Saudi Arabia. Mogadiscio registry closed. Registered at Jeddah as **AL FAYEZ**. 1957: Converted to motor and fitted with a 6-cylinder oil engine by Skoda. Prague. 1958: Sold to G & N Angelakis & Antonios Petras, Piræus, Greece. Jeddah registry closed. Registered at Piræus as **AGHIA MARINA**. 03.03.1963: Stranded at Kamena Vourla, south coast of Malian Gulf, Greece; came afloat under own power three days later. 05.07.1967: On passage from Thessalonika towards Piræus with a cargo of wheat, 40 miles off Cape Doro started to take in water and foundered in position 38.54N 24.28E. All crew saved.

SLUNA	1231		545disp	150.0	Amos & Smith	The Admiralty,
T177	13.09.1940	14.04.1941		27.6	850ihp 3-cyl	Whitehall,
Isles class trawler	03.12.1940	15.10.1941		10.5	12.0 knots	London

Name chosen was SHUNA but in error she was mistakenly launched as SLUNA. Commissioned as a minesweeper (1-12pdr, AA weapons, DC) (P.No.T177). Based Plymouth with 6th Trawler Group. 07.1946: Sold to Chinese interests and renamed **SHUN WHA**. 1948: Sold to Hai Ying Steamship Co Ltd, Shanghai. Converted to dry cargo, 470grt, 146net. Registered at Shanghai as **HAI MA**. 13.10.1950: Stranded in Hainan Strait between the Leizhou Peninsula and Hainan Island, southern China. Salved but beyond economical repair and broken up. 1950: Registry closed.

FRISKY	1232		597disp	142.6	C D Holmes	The Admiralty,
W11	24.09.1940	27.05.1940		33.3	1350ihp 3-cyl	Whitehall,
Assurance class tug	16.01.1941	20.09.1941		11.0	13.0 knots	London

Commissioned (1-3", AA weapons) (P.No.W11). 1948: Sold to Kuwait Oil Co Ltd, London. Registered at London as **HASAN**. Official No.181802. 601grt, 3net. 1961: Sold to Nicolas E Vernicos Shipping Co Ltd, Piræus, Greece. London registry closed. Registered at Piræus as **VERNICOS MARINA**. 1973: Sold to Michael Constantinidis, Piræus. 02.1973: Breaking up commenced at Piræus. 1973: Registry closed.

JAUNTY	1233		597disp	142.3	C D Holmes	The Admiralty,
W30/A140	24.09.1940	11.06.1940		33.3	1350ihp 3-cyl	Whitehall,
Assurance class tug	15.02.1941	06.11.1941		11.0	13.0 knots	London

Commissioned (1-3", AA weapons) (P.No.W30).   12.1941: Based Scapa Flow.   Took part in Operation Anklet, landings on Lofoten Islands.
08.1942: With escorts of convoy WS-21S (Operation Pedestal) to Malta.   11.08.1942: In company with HM Ships LAFOREY (P.No.G99)
(Capt R M J Hutton RN) and LOOKOUT (P.No.G32) (Lt Cdr A G Foreman RN), picked up CO, 485 crew and one news reporter, part of 927
survivors of aircraft carrier EAGLE (Capt Lachlan D Mackintosh RN) torpedoed by U-boat (U73) 70 miles S of Cape Sallnas, Majorca.
EAGLE foundered in position 38.05N 03.02E.   Transferred survivors to HM Ships KEPPEL (P.No.D84), MALCOLM (P.No.D19) and
VENOMOUS (P.No.D75) landing them at Gibraltar.   Later assisted the badly damaged American (British manned) tanker OHIO
(9265grt/1940) to berth in Malta.   06.1944: Operation Neptune - Normandy landings.   05.06.1944: Sailed Solent for Eastern Task Force
Area (follow up convoy L1).   05.06.1944: Connected to LCT2428 which was taking in water but foundered.   07.06.1944: Connected to
hospital carrier ST JULIEN (1952grt/1925) extensively damaged by mine whilst outside swept waters.   08.06.1944: Delivered Southampton.
1948: Decommissioned.   12.1948: Registered at London. Official No.181622. 601grt, 3net.   1949: Recommissioned with Royal Fleet
Auxiliary crew (P.No.A140).   1956-1958: Based Chatham, PAS manned.   1958: Placed in reserve at Pembroke Dock.   28.08.1963: Based
Portland target towing.   19.03.1964: Transferred to Chatham.   15.11.1965: Sold to Jos de Smedt, Antwerp, for breaking up.
21.12.1966: Sold to unidentified owners.   No further details.

EDAY	1234		545disp	150.0	Amos & Smith	The Admiralty,
T201	16.11.1940	10.07.1941		27.6	850ihp 3-cyl	Whitehall,
Isles class trawler	01.04.1941	07.01.1942		10.5	12.0 knots	London

Commissioned as a minesweeper (1-12pdr, AA weapons, DC) (P.No.T201).   Based Scapa Flow/Tobermory with 8th Trawler Group.
08.1944: Transferred on loan to the Royal Norwegian Navy. Renamed TROMØY (P.No.T201).   19.10.1944: Returned and reverted to EDAY
(P.No.T201).   01.1948: Sold to A/S Fjeldøy, Haugesund, Norway.   1948: Converted to dry cargo and fitted with 7-cylinder oil engine by
British Auxiliaries Ltd, Glasgow.   597grt, 245net.   Registered at Haugesund as FJELLBERG.   23.01.1952: Sold to Keller Shipping Ltd,
Basle, Switzerland.   Haugesund registry closed.   Registered at Basle as SEMPACH.   Remeasured to 647grt, 340net and later to 643grt,
309net, 815dwt.   27.04.1953: On a voyage from Nemours, Algeria, towards Tunis with general cargo suffered an explosion and started to
take in water before foundering 5 miles NE of Nemours.   1953: Registry closed.

FETLAR	1235		545disp	150.0	Amos & Smith	The Admiralty,
T202	16.11.1940	10.07.1941		27.6	850ihp 3-cyl	Whitehall,
Isles class trawler	01.04.1941	07.01.1942		10.5	12.0 knots	London

Commissioned as a minesweeper (1-12pdr, AA weapons, DC) (P.No.T202).   1946-1950: Employed as a wreck dispersal vessel (P.No.DV8).
02.06.1960: Sold to Jos de Smedt, Antwerp, Belgium, for breaking up.   22.06.1960: Arrived Antwerp

FOULA	1236		545disp	150.0	Amos & Smith	The Admiralty,
T203	16.11.1940	09.08.1941		27.6	850ihp 3-cyl	Whitehall,
Isles class trawler	05.04.1941	13.02.1942		10.5	12.0 knots	London

Commissioned as a minesweeper (1-12pdr, AA weapons, DC) (P.No.T203).   02.1946: Sold to Italian Navy (Marina Militare) (P.No.RD313).
1965: Expended as a target.

ADEPT	1237		700disp	142.6	C D Holmes	The Admiralty,
W107	10.01.1941	25.08.1941		33.3	1350ihp 3-cyl	Whitehall,
Assurance class tug	29.05.1941	20.02.1942		11.0	13.0 knots	London

Commissioned (1-3", AA weapons) (P.No.W107). Assigned as rescue tug.   17.03.1943: In thick fog stranded on the south-west side of
Paterson's Rock, Mull of Kintyre. Total loss.   2013: Wreck is well dispersed but boiler is in 9m in approximate position 55.17N 05.32W.

ADHERENT	1238		700disp	142.5	C D Holmes	The Admiralty,
W108	10.01.1941	24.09.1941		33.3	1350ihp 3-cyl	Whitehall,
Assurance class tug	12.06.1941	27.03.1942		11.0	13.0 knots	London

Commissioned (1-3", AA weapons) (P.No.W108). Assigned as rescue tug.   14.01.1944: Foundered in North Atlantic with the loss of CO,
Lt Stanley W Potter, RNR and ten crew.

BONITO	1239		670disp	147.8	C D Holmes	The Admiralty,
T231	27.02.1941	08.10.1941		25.1	700ihp 3-cyl	Whitehall,
Fish class trawler	11.07.1941	03.04.1942		12.5	11.0 knots	London

The ten Fish class trawlers were based on the builder's commercial design of GULLFOSS (see yard no.1058) owned by Consolidated
Fisheries Ltd, Grimsby. After the war the eight surviving trawlers were purchased by Consolidated Fisheries Ltd, Grimsby, for an en bloc sum
of £132,000.   Bills of sale dated 24.05.1946.
09.04.1942: Commissioned as an A/S trawler/minesweeper (1-4", AA weapons, DC & ASDIC) (P.No.T231).   24.05.1946: Sold to
Consolidated Fisheries Ltd, Grimsby for the sum of £18,857.   Cost of conversion £29,725 - total £48,582.   04.02.1947: Registered at
Grimsby as BLAEFELL (GY456). Official No.166657. 380grt, 142net, 650ihp. Insured for £60,520.   22.04.1947: Sailed on first trip.
31.12.1954: Sold to Clifton Steam Trawlers Ltd, Fleetwood.   05.01.1955: Grimsby registry closed.   08.01.1955: Registered at Fleetwood
(FD40).   28.06.1956: Company taken over by Boston Deep Sea Fisheries Ltd, Fleetwood.   05.09.1956: Sold to B Gelcer & Co (Proprietary)
Ltd, Cape Town.   03.09.1956: Fleetwood registry closed.   09.1956: Registered at Cape Town (CTA387).   12.1956: Registered at Cape
Town as BENJAMIN GELCER (CTA387).   Pre-1966: Company sold to Irvin & Johnson Ltd, Cape Town, who became managers.
02.1967: Stripped of all usable parts and non-ferrous metal and scuttled off Cape Town.   Cape Town registry closed.

Sluna (1231)

(Cochrane archive)

Foula (1236)

(Cochrane archive)

Bonito (1239)

(Cochrane archive)

WHITING T232 Fish class trawler	1240 27.02.1941 28.07.1941	22.10.1941 07.05.1942	670disp	146.0 25.1 12.5	C D Holmes 700ihp 3-cyl 11.0 knots	The Admiralty, Whitehall, London

Commissioned as an A/S trawler/minesweeper (1-4", AA weapons, DC & ASDIC) (P.No.T232).
24.05.1946: Sold to Consolidated Fisheries Ltd, Grimsby for the sum of £18,858. Cost of conversion £18,695 - total £37,553.
15.10.1946: Registered at Grimsby as **BURFELL** (GY346). Official No.166651. 380grt, 142net, 650ihp. Insured for £60,520.
16.10.1946: Sailed on first trip.   10.03.1958: Sold to Rhondda Fishing Co Ltd, Grimsby.   1960: Sold to BISCO and allocated to Thos Young & Sons Ltd, Sunderland, for breaking up.   10.06.1960: Arrived Sunderland.   05.06.1961: Grimsby registry closed - "Vessel broken up".

CHARON W109 Assurance class tug	1241 26.04.1941 14.08.1941	27.11.1941 11.06.1942	700disp	142.6 33.3 11.0	C D Holmes 1350ihp 3-cyl 13.0 knots	The Admiralty, Whitehall, London

Commissioned (1-3", AA weapons) (P.No.W109). Assigned as rescue tug.   15.08.1944: Took part in Operation Dragoon - Allied landings in Southern France. Task Force 85 Delta Force, Combat & Firefighting Group.   1947: Renamed **ALLIGATOR** (P.No.W109).
1949: Decommissioned.   1954 - 58: Based Portsmouth and Portland under CD.   07.1958: Sold to Henry George Pounds, Paulsgrove, Portsmouth.   12.12.1958: Sold to J D Irving Ltd, Saint John, NB, Canada. Laid up.   1959: Re-engined with T.3-cylinder by J I Thornycroft & Co Ltd, Southampton. Registered at Southampton as **IRVING BIRCH**. Official No.300642. 616grt.   11.01.1960: Attended coaster AHERN TRADER (774grt/1922) aground after leaving wharf at Fredericton, Notre Dame Bay, Newfoundland. Connected but failed to refloat. Engaged to conduct salvage before wreck abandoned.   03.1963: Sailed Dartmouth, NS, with wooden motor/sail vessel BEAR (648grt/1874) in tow bound Philadelphia.   11.03.1963: In a severe gale, about 100 miles E of Cape Sable, NS, tow parted, mast fell puncturing hull, started to take in water and foundered; crew picked up by tug.   12.1964: Laid up.   1967: Registered at Southampton as **IRVING FORTY** to free name for new build.   Prior to 11.1969: Foundered at Indiantown, NB. Refloated.   26.11.1969: Sold to Abe Levine & Sons Ltd, Fredericton, NB. Being dismantled at Indiantown, NB; possibly sold.   12.05.1970: Sank at moorings at Indiantown, NB. Refloated and dismantling continued.   Circa 1972: Hulk towed out to sea and scuttled.   1972: Registry closed.

DECISION W110/A110 Assurance class tug	1242 1941 27.08.1941	20.12.1941 23.07.1942	700disp	142.6 33.3 11.0	C D Holmes 1350ihp 3-cyl 13.0 knots	The Admiralty, Whitehall, London

Commissioned (1-3", AA weapons) (P.No.W110). Assigned as rescue tug. Based Trincomalee, Ceylon, with East Indies fleet.
1947: Paid off.   1948: Renamed **HENGIST** (P.No.A110). Royal Fleet Auxiliary manned.   1954: At Pembroke Dock in reserve.
15.07.1964: Sold to D Arnold, Ashford, Middlesex. Towed to Cardiff and laid up.   12.1964: Sold to Tsavliris (Salvage & Towage) Ltd, Piræus, Greece.   1965: Registered at Piræus as **NISOS CRETE**. 597grt, 1net.   03.1973: Sold to Papageorgiou Salvage & Towage Ltd, Greece, and broken up.   Registry closed.

EMPIRE FAIRY 168783 Hoedic class tug	1243 1941 1941	05.01.1942 19.05.1942	277	105.3 26.6 12.2	Amos & Smith 825ihp 3-cyl 11.0 knots	Ministry of War Transport, London

The eighteen Hoedic class tugs were based on the builder's commercial design of HOEDIC (see yard no.1105).
05.1942: Registered at Goole.   30.09.1942: Delivered to Mombasa by Pedder & Mylchreest Ltd, London, and placed under management of Kenya & Uganda Railways.   12.01.1944: Taken over by the Admiralty and based at Kilindini Harbour with the Eastern Fleet.
01.01.1946: Sold to Rangoon Port Commissioners, Rangoon, Burma.   1948: Goole registry closed. Registered at Rangoon as **NATHAMEE**.   1954: Owners restyled Board of Management for the Port of Rangoon.   1972: Owner restyled Burma Ports Corporation.
1989: Owner restyled Myanma Port Authority, Yangon, Myanmar.   2015: No further trace.

EMPIRE GOBLIN 168784 Hoedic class tug	1244 1941 1941	19.01.1942 05.06.1942	277	105.3 26.6 12.2	Amos & Smith 825ihp 3-cyl 11.0 knots	Ministry of War Transport, London

06.1942: Registered at Goole.   30.09.1942: Delivered to Cape Town by Pedder & Mylchreest Ltd, London, and placed under management of the Government of the Union of South Africa as a search and rescue tug patrolling between Cape Town and Dassen Island.
04.1946: Transferred to Ministry of Transport, London.   1947: Employed as a harbour tug at Cape Town.   1948: Sold to Cia. Argentina de Nav Angel, Gardella Ltda, Buenos Aires, Argentina. Goole registry closed. Registered at Buenos Aires as **BIO BIO**.   10.1992: Sold to Maruba S.C.A. Empresa de Navegacion Maritima, Panama. Buenos Aires registry closed. Registered at Panama. Operating in the port of Quequén-Necochea and later Sante Fe.   1990s: Laid up in port at Sante Fe along with steam tug TRIUNFADOR (213grt/1933).
01.2013: Both surveyed for possible preservation by Histamar Foundation, Buenos Aires.

GRAYLING T243 Fish class trawler	1245 13.06.1941 23.10.1941	04.03.1942 02.07.1942	670disp	146.0 25.3 12.5	Amos & Smith 700ihp 3-cyl 11.0 knots	The Admiralty, Whitehall, London

Commissioned as an A/S trawler/minesweeper (1-4", AA weapons, DC & ASDIC) (P.No.T243).
24.05.1946: Sold to Consolidated Fisheries Ltd, Grimsby, for the sum of £18,857. Cost of conversion £26,818 - total £45,675.
02.1947: Registered at Swansea as **BARRY CASTLE** (SA33). Official No.168574. 380grt, 142net. Insured for £60,520.
10.03.1947: Sailed on first trip out of Swansea.   11.1953: Transferred to Grimsby.   01.11.1955: On an Icelandic trip (Sk Walter Oxley); eighteen crew. Off West Fjords in very heavy weather and Force 10 storm, started to take in water through cracked bunker lid. With engine room water level rising, motor trawler PRINCESS ELIZABETH (H238) (see yard no.1380) connected and commenced tow towards Isafjordur. Started to settle and before foundering ten crew were rescued by trawler VIVIANA (GY233) (Sk James Gamble) (see yard no.1157), two crew by trawler STAFNES (GY172) (see yard no.1141) and two by trawler CAPE PORTLAND (H357) (see yard no.1170). Four crew (Ch Eng, 2nd Eng, Bosun and a deckhand) lost. All transferred to STAFNES and survivors landed at Isafjordur for transfer to HMS ROMOLA (P.No.J449) for passage to Reykjavik.   1956: Swansea registry closed. The ship's bell was later brought up by an Icelandic trawler, returned to UK and is now in the National Fishing Heritage Centre, Grimsby.

MACKEREL	1246		670disp	146.5	Amos & Smith	The Admiralty,
M55	13.06.1941	06.03.1942		25.0	700ihp 3-cyl	Whitehall,
Fish class trawler	10.1941	07.12.1942		12.5	11.0 knots	London

Renamed **CORNCRAKE** (P.No.M55). Completed as a controlled mine-layer (AA Weapons, mines (12)).   25.01.1943: Foundered in heavy weather in the North Atlantic. Despite search by sloop HMS LONDONDERRY (P.No.L76), twenty-three crew lost including CO, Ty/Act/Lt Cdr Lewis R Renfrew RNR.

DEXTEROUS	1247		700disp	142.6	C D Holmes	The Admiralty,
W111	01.07.1941	03.04.1942		33.3	1350ihp 3-cyl	Whitehall,
Assurance class tug	26.11.1941	10.09.1942		11.0	13.0 knots	London

Commissioned (1-3", AA weapons) (P.No.W111). Assigned as rescue tug.   01.1943: Transferred to the Ministry of War Transport. 03.1943: Chartered by L Smit & Co's Internationale Sleepdienst, Rotterdam. Manned by Dutch crew (Capt Kalkman). Registered at Rotterdam. Based St John's, Newfoundland, as convoy escort/rescue tug.   10.05.1943: Part of Escort Group C2. Became separated in fog and attacked by U-boat (U403) with gunfire which was returned. Informed Convoy HX237 that under attack and Swordfish from HMS BITER strafed and dropped two bombs on the U-boat which escaped without damage.   06.1944: Operation Neptune - Normandy landings. Assigned to Senior Naval Officer (SNO) Selsey.   11.06.1944: Towing Whale units to Mulberry A and B.   07.1945: Returned to Ministry of War Transport. Registered at Leith. Official No.167032; 600grt, 45net.   23.11.1945: Sold to Overseas Towage & Salvage Ltd, London. Leith registry closed. Registered at London.   18.09.1951: Sailed Rio de Janeiro with BUSTLER towing Brazilian battleship SAO PAULO to UK for breaking up.   04.11.1951: Off Azores suffered heavy weather damage.   1957: Sold to BP Tanker Co Ltd, London. Registered at London as **ZURMAND**.   1968: Sold to Tsavliris (Salvage & Towing) Ltd, Piræus. London registry closed. Registered at Piræus as **NISOS IKARIA**.   1969: Sold to shipbreakers.   1971: Broken up at Perama.   Registry closed.

GRIPER	1248		700disp	142.6	C D Holmes	The Admiralty,
W112	01.07.1941	16.05.1942		33.3	1350ihp 3-cyl	Whitehall,
Assurance class tug	23.12.1941	06.10.1942		11.0	13.0 knots	London

Commissioned (1-3", AA weapons) (P.No.W112). Assigned as rescue tug.   31.08.1943: Arrived Sydney, Nova Scotia, with burning American 'Liberty' ship J PINKNEY HENDERSON (7176grt/1943) (Capt Clarence H Lundy) in tow having been involved in collision off Newfoundland with American tanker J H SENIOR (11065grt/1931) loaded with avgas, both vessels engulfed in flames.   20.04.1945: Allocated for tropicalisation by Thames shiprepairers.   1946: Based Singapore.   27.12.1946: With tug ASSIDUOUS (P.No.W142) (see yard no.1269) towed Japanese cruiser TAKAO from Selatar to Malacca Straits prior to scuttling.   1949: Sold to Singapore Harbour Board, Singapore. Registered at Singapore. Official No.179938;. 597grt, 21net.   1962: Sold to Ta Hing Co (HK) Ltd, Panama. Singapore registry closed. Registered at Panama as **SURABAJA**.   09.05.1962: Sailed Singapore Dockyard with damaged destroyer HOGUE (ex P.No.D74) in tow for breakers.   1962: Sold to Republik Indonesia Perusahaan Negara Tundabara, Djakarta, Indonesia. Panama registry closed. Registered at Djakarta as **SELAT SURABAJA**.   1970: Sold to P N Pelajaran Bahtera Adhiguna, Djakarta, and broken up in Indonesia. Registry closed.

EMPIRE PAT	1249		274	105.3	Amos & Smith	Ministry of War Transport,
168788	1941	30.05.1942		26.6	825ihp 3-cyl	London
Hoedic class tug	1941	26.08.1942		12.2	11.0 knots	

1942: Registered at Goole. Delivered to Persian Gulf by Pedder & Mylchreest Ltd, London, and employed on Admiralty service with East Indies Command, based Basra.   1946: Released from Admiralty service and transferred to Ministry of Transport.   1946: Chartered to Anglo Iranian Oil Co Ltd, London, and based at Bandar-e Mahshahr, Iran.   1949: Sold to Kuwait Oil Co Ltd, Kuwait. Goole registry closed. Registered at London as **HIMMA**.   01.1951: Sold to J Fenwick & Co Pty Ltd, Sydney, Australia. London registry closed. Registered at Sydney, NSW.   07.1960: Based at Port Kembla, NSW.   02.1971: Laid up in Morts Bay, Sydney.   1972: Sold to Pimco Shipping Co Pty Ltd, Port Moresby, Papua New Guinea. Sydney registry closed. Registered at Port Moresby.   Conversion work started to fit out for cargo to operate around Papua New Guinea but damaged during conversion by a berthing container ship; work stopped.   1974: Sold to W J Byers, Port Moresby, and remained in Morts Bay.   Owner died and repossessed by J Fenwick & Co Pty Ltd.   Stripped of much non ferrous metals. 30.08.1977: Scuttled in 48m of water, along with nine other vessels to form an artificial reef to attract fish and to create a recreational dive site, at 'Ship Reef 'off Narrabeen Beach, North Sydney Heads.   1978: Registry closed.

EMPIRE SAM	1250		274	105.3	Amos & Smith	Ministry of War Transport,
168789	1941	01.06.1942		26.6	825ihp 3-cyl	London
Hoedic class tug	1942	07.09.1942		12.2	11.0 knots	

09.1942: Registered at Goole.   09.1942: Delivered to Pedder & Mylchreest Ltd, London, for delivery to Colombo, Ceylon, and employed on Admiralty service with East Indies Command.   03.04.1945: Sailed Colombo for Darwen, Northern Territory, Australia.   12.06.1945: Arrived Darwin.   30.08.1945: Sailed Darwin for Hong Kong.   18.09.1945: Arrived Hong Kong and employed as Dockyard tug.   1949: Released from Admiralty service and transferred to Ministry of Transport..   1949: Sold to the Hong Kong Government, Hong Kong, for use by Hong Kong Police. Goole registry closed.   1965: Sold to Yau Wing Co Ltd, Hong Kong. Registered at Hong Kong as **YAU WING No.23**.   1966: Sold to Cheung Chau Shipping & Trading Co, Hong Kong.   1966: Sold to Pan Asiatic Lines Inc of Carolus SA, Panama. Hong Kong registry closed. Registered at Panama as **FEDREDGE SAM**.   1967: Sold to Wo Hing Co, Hong Kong, for breaking up.   09.1967: Commenced breaking up. 1967: Registry closed.

PROSPEROUS	1251		700disp	142.6	C D Holmes	The Admiralty,
W96	29.10.1941	29.06.1942		33.3	1350ihp 3-cyl	Whitehall,
Assurance class tug	09.03.1942	18.11.1942		11.0	13.0 knots	London

Commissioned (1-3", AA weapons) (P.No.W96). Assigned as rescue tug. Convoy escort duties.   22.01.1944: Took part in Operation Shingle - Italian Campaign, Anzio landings.   1945: Took part in Operation Deadlight - towage and scuttling of surrendered German U-boats. 05.1947: Decommissioned. Based Portsmouth as Dockyard tug.   1949: Based Portsmouth manned by Royal Fleet Auxiliary. Registered at Harwich. Official No.182198; 597grt, 2net. Engaged in ocean and coastal towage.   12.1960: On completion of refit at Portsmouth, placed in Reserve.   05.1961: Placed in Reserve at Chatham Dockyard.   11.1964: For sale at Chatham.   07.01.1965: Sold to Aegean Steam Navigation Typaldos Bros Ltd, Piræus, Greece. Harwich registry closed. Registered at Piræus as **EYFORIA**. Remeasured to 626grt. 1967: Sold to Varnima Corp, Piræus. Registered at Piræus as **EFORIA**. 1968: Sold to Seka S.A, Rethymnon.   Registered at Piræus as **CAPTAIN SPYROMILIOS**. Remeasured to 555grt and employed as a pilot vessel.   1980: Sold to United Shipbreakers Ltd, Eleusis, for breaking up.   23.08.1980: Arrived Eleusis under tow.   1980: Registry closed.

RESCUE	1252		700disp	142.5	C D Holmes	The Admiralty,
W97	1942	07.1942		33.0	1350ihp 3-cyl	Whitehall,
Assurance class tug	04.1942	17.12.1942		11.0	13.0 knots	London

Commissioned (1-3", AA weapons) (P.No.W97). 1942: Renamed **HORSA**. 1943: Based Iceland as rescue tug/escort.
16.03.1943: Stranded on the east coast of Iceland near Osfles Rock. Total loss.

MULLET	1253		670disp	147.8	Amos & Smith	The Admiralty,
T311	17.12.1941	14.08.1942		25.1	700ihp 3-cyl	Whitehall,
Fish class trawler	14.05.1942	12.11.1942		12.5	11.0 knots	London

Commissioned as an A/S trawler/minesweeper (1-4", AA weapons, DC & ASDIC) (P.No.T311).
24.05.1946: Sold to Consolidated Fisheries Ltd, Grimsby for the sum of £18,857. Cost of conversion £26,967 - total £45,824.
02.1947: Registered at Swansea as **NEATH CASTLE** (SA49). Official No.168577; 380grt, 142net. Insured for £60,520. 04.06.1947: Sailed on first trip out of Swansea. 03.1958: Sold to Rhondda Fishing Co Ltd, Grimsby. 03.1958: Swansea registry closed.
20.03.1958: Registered at Grimsby (GY52). 10.06.1960: Sold to BISCO and allocated to Clayton & Davie Ltd, Dunston, for breaking up.
06.1960: Arrived River Tyne. 05.06.1961: Grimsby registry closed.

TURBOT	1254		670disp	146.0	Amos & Smith	The Admiralty,
M31	17.12.1941	28.08.1942		25.3	700ihp 3-cyl	Whitehall,
Fish class trawler	18.05.1942	06.01.1943		12.5	11.0 knots	London

Renamed **REDSHANK** (P.No.M31). Completed as a controlled mine-layer (AA Weapons, mines (12)). 1957: Decommissioned for disposal. 04.1957: Sold to BISCO and allocated to Thomas Young & Sons (Shipbreakers) Ltd, Dunston, for breaking up.
09.05.1957: Arrived South Dock, Sunderland.

EMPIRE ACE	1255		275	105.3	C D Holmes	Ministry of War Transport,
169078	1942	12.09.1942		26.6	825ihp 3-cyl	London
Hoedic class tug	1942	22.12.1942		12.2	11.0 knots	

12.1942: Registered at Goole. Delivered to Pedder & Mylchreest Ltd, London, for delivery to Malta. 15.03.1944: Sunk during an air attack on Malta. 10.05.1944: Salved and repaired. 05.03.1946: Civilian manned at Malta under Captain of the Dockyard.
13.03.1947: Transferred to the Admiralty. 08.1947: Goole registry closed. 29.08.1947: Renamed **DILIGENT** (unregistered).
18.09.1958: Sailed Malta for Devonport in tow of SAUCY (see yard no.1257). 1960: In reserve at Pembroke Dock. 1961: Transferred to the Secretary of State for Defence, London. 27.04.1961: Registered at London as **EMPIRE ACE**. 04.1961: Loaned to the United States Navy for operation at the Holy Loch facility and River Clyde. 31.12.1964: Returned to Secretary of State for Defence, London. Royal Fleet Auxiliary manned under CAPIC Clyde operational control. 11.11.1968: Approaching Campbeltown in heavy seas (Capt E E Shelton), driven ashore on Mull of Kintyre four miles south of Campbeltown. Crew took to boats and landed safely. Because of poor weather all attempts to refloat failed and abandoned. 28.03.1969: Refloated by MSBV MANDARIN and MFV64 and delivered Campbeltown. 18.06.1969: Survey revealed extensive damage and declared beyond economical repair. 26.06.1969: Offered for sale. 06.1969: London registry closed.
30.09.1969: Sold to George Hood, Helensburgh. 1970: Sold to Archibald Macfadyen, Campbeltown, for breaking up. 04.1971: Stripped of all non-ferrous metals and superstructure cut down.

EMPIRE DENNIS	1256		274	105.3	C D Holmes	Ministry of War Transport,
169079	1942	26.09.1942		26.6	825ihp 3-cyl	London
Hoedic class tug	1942	13.01.1943		12.2	11.0 knots	

01.1943: Registered at Goole. Delivered to Pedder & Mylchreest Ltd, London, for delivery to Malta. Allocated to CinC Mediterranean.
1945: Delivered Naples by Pedder & Mylchreest Ltd, London, for service at Naples with Italian crew. 04.1946: Transferred to Ministry of Transport, London. 06.1946: Released from naval service. 27.08.1946: Placed in C&M in Malta pending return to UK. 1947: Sold to Clyde Shipping Co Ltd, Glasgow. 1947: Goole registry closed. Registered at Glasgow. 1948: Registered at Glasgow as **FLYING METEOR**. 19.08.1962: Sold to I C Guy Ltd, Cardiff. Glasgow registry closed. 1962: Registered at Cardiff as **ROYAL ROSE**.
30.06.1963: Company in voluntary liquidation; company and ships sold to R & J H Rea Ltd, London. Registered at Cardiff as **YEWGARTH**.
14.09.1965: Crushed against lock wall and holed whilst assisting motor ore carrier ALDERSGATE (12,718grt/1960) to lock into Cardiff Docks. Making water, taken in tow by motor tug TREGARTH (102grt/1958) and beached outside the dock entrance. 20.09.1965: Salved but declared a constructive total loss. Sold to John Cashmore Ltd, Newport, for breaking up. 21.09.1965: Delivered Newport under tow.
30.09.1965: Breaking commenced. 1965: Registry closed.

SAUCY	1257		700disp	142.6	C D Holmes	The Admiralty,
W131/A386	21.01.1942	26.10.1942		33.3	1350ihp 3-cyl	Whitehall,
Assurance class tug	30.06.1942	25.02.1943		11.0	11.0 knots	London

Commissioned (1-3", AA weapons) (P.No.W131). Assigned as rescue tug. Convoy escort duties. 06.1944: Operation Neptune - Normandy landings. Assigned to Senior Naval Officer (SNO) Selsey. 08.06.1944: Towing Whale units to Mulberry B. 27.06.1944: Whale tow foundered. 1947: Registered at Harwich. Official No.182196. 597grt, 1net. Laid up. 1949: Royal Fleet Auxiliary manned (P.No.A386). Based Portsmouth/Portland target towing duties. 1955: Based Malta. 18.09.1958: Sailed Malta for Devonport towing EMPIRE ACE (see yard no.1255). 11.11.1960: Placed in operational reserve at Pembroke Dock. 01.03.1965: Sold to Tsvliris (Salvage & Towage) Ltd, Piræus, Greece. Harwich registry closed. Registered at Piræus as **NISOS CHIOS**. Remeasured to 555grt. 1973: Sold to shipbreakers in Greece and broken up. 1973: Registry closed.

Charon (1241), an Assurance class tug.

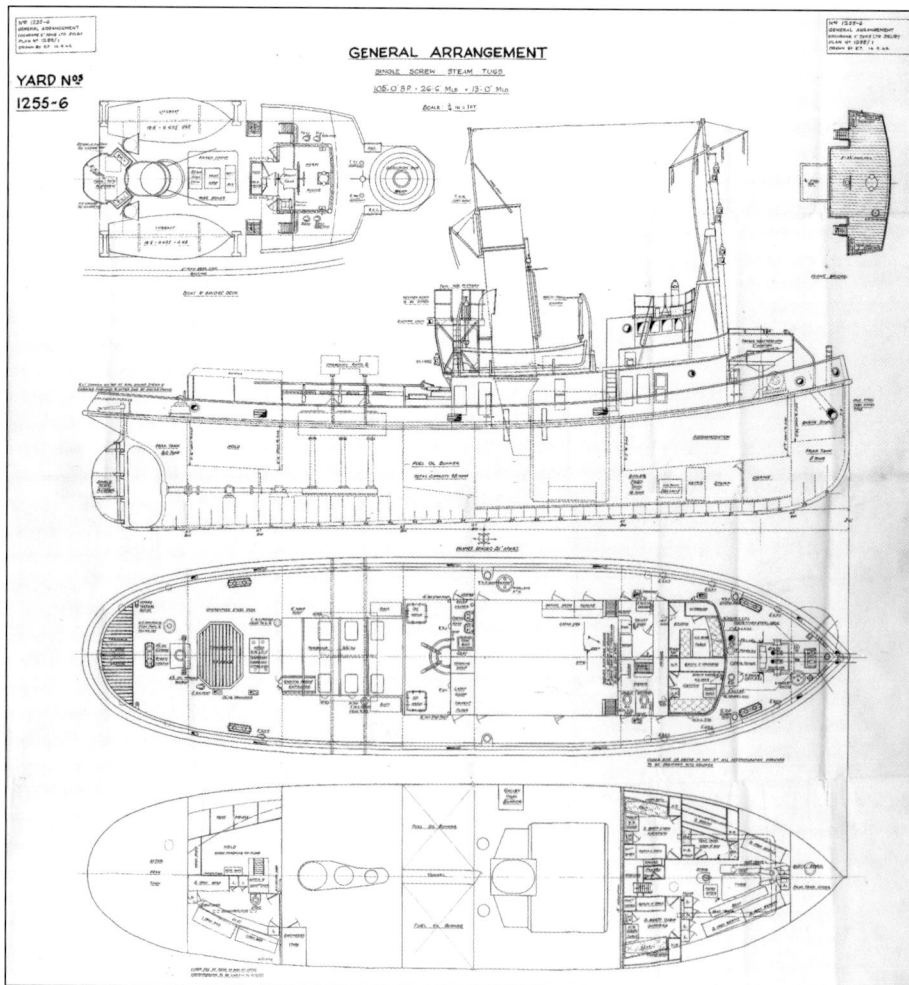

General arrangement drawing for **Empire Ace** (1255) and **Empire Dennis** (1256), Hoedic class tugs.

We have two fine images of the Hoedic-class tug **Empire Dennis** (1256) in her later career. Firstly we see her at Barry as **Royal Rose**.

(Stuart Emery collection)

We now see her as she entered the port of Barry as the **Yewgarth** of R & J H Rea.

(Stuart Emery collection)

STORMCOCK	1258		700disp	142.6	C D Holmes	The Admiralty,
W87	21.01.1942	24.11.1942		33.3	1350ihp 3-cyl	Whitehall,
Assurance class tug	30.07.1942	25.03.1943		11.0	11.0 knots	London

1943: Commissioned as **STORMKING** (1-3", AA weapons) (P.No.W87). Assigned as rescue tug. 06.1944: Operation Neptune - Normandy landings. Assigned to Senior Naval Officer (SNO) Selsey. 11.06.1944: Towing Whale units to Mulberry B. 11.06.1944: Picked up survivors from HMRT SESAME (see yard no.1275) torpedoed and sunk by German e-boat off Normandy. 1946: De-commissioned. 1947: Renamed **TRYPHON** (unregistered). Based Sheerness/Chatham. 07.1957: Sold to Foremost Marine Transporters Ltd, London. 1959: Registered at London as **MELANIE FAIR**. Official No.301044. 611grt. 09.05.1960: Foundered in St Lawrence river, 30 miles west of Quebec, after striking a submerged obstruction. 12.05.1960: Salved and repaired. 03.08.1960: Sailed Norfolk, VA (Capt W Wharton) for Antwerp towing after part of Liberian tanker AFRICAN QUEEN (13,759grt/1955) sold for breaking up. 1961: Sold to Imprese Marittime e Portuali S.p.A, Genoa, Italy. London registry closed. Registered at Genoa as **TORO**. Remeasured to 588grt. 04.1969: Sold to shipbreakers in Italy and broken up at La Spezia. 1969: Registry closed.

BREAM	1259		670disp	146.0	Amos & Smith	The Admiralty,
T306	17.04.1942	10.12.1942		25.3	700ihp 3-cyl	Whitehall,
Fish class trawler	31.08.1942	27.04.1943		12.5	11.0 knots	London

Commissioned as an A/S trawler (1-4", AA weapons, DC & ASDIC) (P.No.T306). Russian convoy duties. 24.05.1946: Sold to Consolidated Fisheries Ltd, Grimsby, for the sum of £18,857. Cost of conversion £19,022 - total £37,879. 26.11.1946: Registered at Grimsby as **VALAFELL** (GY383). Official No.166654. 380grt,142net. 27.12.1946: Sailed on first trip. 20.07.1955: Arrested for illegal fishing in Thistilfjorður, Iceland. Skipper A W Bruce fined 74,000 kronur (£1619); gear and catch confiscated. 10.03.1958: Sold to Rhondda Fishing Co Ltd, Grimsby. 06.1960: Sold to BISCO and allocated to C W Dorkin & Co Ltd, Dunston. 10.06.1960: Delivered River Tyne but broken up by Clayton & Davie Ltd at Dunston. 05.06.1961: Grimsby registry closed - "Vessel broken up".

Bream (1259) in her original guise as a Fish class trawler.

(Cochrane archive)

After World War II, the above vessel was converted to a conventional trawler and renamed **Valafell***.*

(Jonathan Grobler collection)

HERRING T307 Fish class trawler	1260 17.04.1942 01.09.1942	24.12.1942 13.04.1943	670disp	146.0 25.3 12.5	Amos & Smith 700ihp 3-cyl 11.0 knots	The Admiralty, Whitehall, London

Commissioned as an A/S trawler (1-4", AA weapons, DC & ASDIC) (P.No.T307). 22.04.1943: Escorting east coast convoy FN108, in collision with French steamer CASSARD (1596grt/1920) in Druridge Bay, 2nm north of 20E buoy NE of Blyth and foundered in position 55.17N 01.19W. All crew picked up.

EMPIRE DARBY 169083 Non-standard Hornby class tug	1261 1942 1942	08.01.1943 27.04.1943	203	95.2 25.1 12.4	McKie & Baxter 750ihp 3-cyl 11.0 knots	Ministry of War Transport, London

04.1943: Registered at Goole. Employed on miscellaneous naval duties. 05.1944: Employed on coastal towage and miscellaneous naval duties. 07.02.1946: Based Chatham on port duties. 04.04.1947: Transferred to the Admiralty. Based Chatham. 08.1947: Goole registry closed. 29.08.1947: Renamed **EGERTON** (unregistered). (Recorded later still as EMPIRE DARBY (unregistered)). 31.12.1957: Following inclining, stability problems and declared unseaworthy. Laid up for disposal. 29.07.1958: Sold to H G Pounds, Portsmouth. 1961: Sold to J D Irving Ltd, Saint John, New Brunswick, Canada. Registered at Saint John, NB as **IRVING BEECH**. 1962: Converted to motor and fitted with 16-cylinder oil engine by General Motors Corp, Cleveland, Ohio. 01.12.1967: Off coast of Nova Scotia, experienced machinery problems whilst towing tanker LUBROLAKE (1622grt/1937) and a barge and all three vessels drifted ashore near New Waterford at the entrance to Sydney Harbour. All abandoned and lost. 1968: Registry closed.

EMPIRE JOAN 169084 Non-standard Hornby class tug	1262 1942 1942	08.01.1943 01.05.1943	203	95.2 25.1 12.4	McKie & Baxter 750ihp 3-cyl 11.0 knots	Ministry of War Transport, London

04.1943: Registered at Goole. Employed on coastal towage. 25.09.1943: Employed on miscellaneous naval duties. 08.10.1945: Employed at Dover. 04.07.1946: Transferred to the Admiralty. Based Chatham; dock work. 08.1947: Goole registry closed. Renamed **EMPHATIC** (unregistered). (Recorded later as EMPIRE JOAN (unregistered)). 01.1958: Following inclining, stability problems and declared unseaworthy. Restrictions placed on harbour use until replacement tug allocated. Placed on sale list. 23.07.1958: Sold to Edmund Hancock (1929) Ltd, Cardiff. Registered at Cardiff. 1960: Following restructuring of towing companies in Bristol Channel, owners became Bristol Channel Towage Co Ltd, Cardiff. 1962: Owners re-styled Bristol Channel Tugs Ltd, Cardiff. 30.06.1963: Company in voluntary liquidation; company and ships sold to R & J H Rea Ltd, London. Later registered at Cardiff as **HALLGARTH**. 1966: Sold to John Cashmore Ltd, Newport, for breaking up. 24.05.1966: Arrived Newport. 1966: Registry closed.

ALLEGIANCE W50 Assurance class tug	1263 1942 27.10.1942	22.02.1942 20.05.1943	700disp	142.6 33.3 11.0	C D Holmes 1350ihp 3-cyl 11.0 knots	The Admiralty, Whitehall, London

Commissioned as a rescue tug (1-3", AA weapons) (P.No.W50). 06.1944: Operation Neptune - Normandy landings. Assigned to Senior Naval Officer (SNO) Selsey. 08.06.1944: Towing Whale units to Mulberry B. 12.1944: Towed floating crane MOWT No.13, Southampton to Arromanches and returned to Southampton with floating crane MOWT No. 11. 1949: Transferred to Hong Kong. Registered at Hong Kong as **ALLEGIANCE 2**. Official No.191430. 597grt. Royal Fleet Auxiliary manned. 1955: Chartered by Hong Kong & Whampoa Dock Co Ltd, Kowloon, Hong Kong. Registered at Hong Kong as **KOWLOONDOCKS**. 30.08.1962: Typhoon Wanda. Towing Norwegian steamer SLETHOLM (3576grt/1950) from Shanghai towards Hong Kong, cast off tow to ride out typhoon. 01.09.1962: Overwhelmed by stress of weather and foundered some 100 miles off Hong Kong. One survivor of crew of thirty. 1962: Registry closed.

ANT W141/A141 Assurance class tug	1264 1942 25.11.1942	24.03.1943 25.06.1943	597	142.6 33.3 11.0	C D Holmes 1350ihp 3-cyl 11.0 knots	The Admiralty, Whitehall, London

Ordered as ANT but name changed before completion. 11.06.1943: Registered at Hull as **ANTIC**. Official No.169288. Commissioned as a rescue tug (1-3", AA weapons) (P.No.W141). Royal Fleet Auxiliary manned. Allocated for Atlantic convoy duties. 1943: Loaned to Royal Netherlands Navy. 21.03.1944: Transferred to Ministry of War Transport, London. 1945: Returned. 17.02.1947: Transferred to Ministry of Transport, London. 26.01.1948: Commissioned as a tender to HMS EXCELLENT, Whale Island, Portsmouth; RN manned (P.No.A141). 19.04.1948: Hull registry closed. 1956: Decommissioned, transferred to CD Portsmouth. 1959: Transferred to Portland. 1968: Transferred to Rosyth. 12.11.1969: Sold to Hughes Bolckow Ltd, Blyth, for breaking up. 14.11.1969: Sailed Rosyth for Blyth in tow of motor tug IRONSIDER (156grt/1967).

GRILSE T368 Fish class trawler	1265 02.06.1942 24.12.1942	06.04.1943 02.07.1943	670disp	146.0 25.3 13.2	Amos & Smith 700ihp 3-cyl 11.0 knots	The Admiralty, Whitehall, London

Commissioned as an A/S trawler (1-4", AA weapons, DC & ASDIC) (P.No.T368). Based Mediterranean. 24.05.1946: Sold to Consolidated Fisheries Ltd, Grimsby, for the sum of £18,857. Cost of conversion £26,239 - total £45,096. 06.1946: Registered at Swansea as **CARDIFF CASTLE** (SA66). Official No.168578; 380grt, 142net. Insured for £75,680. 06.1947: Fitted for burning oil fuel. 24.06.1947: Sailed on first trip. 1952: Sold to Clifton Steam Trawlers Ltd, Fleetwood. 05.1952: Swansea registry closed. 08.05.1952: Registered at Fleetwood (FD103). 13.07.1952: Registered at Fleetwood as **JULIA BRIERLEY** (FD103). 06.07.1956: Company taken over by Boston Deep Sea Fisheries Ltd, Fleetwood. 04.04.1957: Sold to Carry On Fishing Co Ltd, Fleetwood. 20.08.1958: Sold to Boston Deep Sea Fisheries Ltd, Fleetwood. 10.1961: Sold to Ets van Heygen Frères SA, Brugge, Belgium, for breaking up. 10.10.1961: Fleetwood registry closed.

POLLOCK T347 Fish class trawler	1266 02.06.1942 28.12.1942	22.04.1943 22.07.1943	670disp	146.0 25.3 13.2	Amos & Smith 700ihp 3-cyl 11.0 knots	The Admiralty, Whitehall, London

Commissioned as an A/S trawler (1-4", AA weapons, DC & ASDIC) (P.No.T347). 24.05.1946: Sold to Consolidated Fisheries Ltd, Grimsby for the sum of £18,857. Cost of conversion £24,934 - total £43,791. Registered at Swansea as **SWANSEA CASTLE** (SA27). Official No.168576; 380grt, 142net. Insured for £61,680. 30.08.1947: Sailed on first trip. 03.1958: Sold to Rhondda Steam Fishing Co Ltd, Grimsby. 03.1958: Swansea registry closed. 20.03.1958: Registered at Grimsby (GY51). 06.1960: Sold to BISCO and allocated to Thomas Young & Sons (Shipbreakers) Ltd, Sunderland, for breaking up. 10.06.1960: Delivered South Docks, Sunderland. 28.08.1961: Grimsby registry closed.

EMPIRE SARA 169307 Hoedic class tug	1267 1942 1942	06.05.1943 12.08.1943	276	105.2 26.6 12.2	Amos & Smith 850ihp 3-cyl 11.5 knots	Ministry of War Transport, London

07.08.1943: Registered at Hull. Employed on port work on River Tyne. 02.1944: Employed on coastal towage. 23.05.1944: With tug EUSTON CROSS (226grt/1924) sailed Tees towing Phoenix units bound Dungeness. 06.1944: Operation Neptune - Normandy landings. Assigned to Senior Naval Officer (SNO) Selsey. 16.06.1944: Arrived Mulberry A towing Whale unit 502/M. 14.09.1944: Employed on miscellaneous naval duties. 22.05.1946: Sold to Ellerman's Wilson Line Ltd, Hull, for £22,500. 01.07.1946: Registered at Hull as **PRESTO**. 13.03.1968: Sold to United Towing Co Ltd, Hull. Laid up. 15.06.1968: Sold to Hughes Bolckow Ltd, Blyth, for breaking up. 24.06.1968: Arrived Blyth under tow along with FORTO (180grt/1939) also for breaking up. 24.06.1968: Hull registry closed.

EMPIRE SIBYL 169313 Hoedic class tug	1268 1942 1942	07.05.1943 26.08.1943	276	105.2 26.6 12.2	Amos & Smith 850ihp 3-cyl 11.5 knots	Ministry of War Transport, London

23.08.1943: Registered at Hull. Employed on port work on River Tyne. 11.05.1946: Sold to Mersey Docks & Harbour Board, Liverpool, for £21,000. 17.06.1946: Hull registry closed. 06.1946: Registered at Liverpool as **ASSISTANT**. 01.1947: Fitted for burning fuel oil. 1962: Sold to Alexandra Towing Co Ltd, Liverpool. Registered at Liverpool as **CASWELL**. Based Swansea. 1969: Sold to Haulbowline Industries Ltd, Passage West, County Cork, for breaking up. 25.03.1969: Arrived Passage West under tow from Swansea. 1969: Registry closed

ASSIDUOUS W142 Assurance class tug	1269 1943 23.02.1943	04.06.1943 23.02.1943	597	156.8 33.2 11.0	C D Holmes 1350ihp 3-cyl 13.0 knots	The Admiralty, Whitehall, London

23.09.1943: Registered at Hull as **ASSIDUOUS**. Official No.169321. Commissioned as a rescue tug (1-3", AA weapons) (P.No.W142). Allocated to Atlantic convoy duties. 06.1944: Operation Neptune - Normandy landings. Assigned to Senior Naval Officer (SNO) Selsey. 07.06.1944: Fouled propeller with tow wire. 11.06.1944: Towing Whale units to Mulberry A and B. 05.07.1944: Sailed Seine Bay with convoy FTC27 towing British coaster WESTDALE (424grt/1911) which had been salved following mine damage and beaching. 1946: Transferred to CD Singapore. 14.02.1946: Towed Japanese submarine I-502 from Selatar to Malacca Straits prior to scuttling. 27.12.1946: With tug GRIPER (P.No.W112) (see yard no.1248) towed Japanese cruiser TAKAO from Selatar to Malacca Straits prior to scuttling. 1947: Commissioned and based Trincomalee, Ceylon; CinC East Indies, RN manned. 1953: Based Gibraltar. 1957: Decommissioned and transferred to CD Gibraltar. 02.07.1958: Laid up for disposal. 10.09.1958: Sold to Henry George Pounds, Paulsgrove, Portsmouth. 12.12.1958: Sold to J D Irvine Ltd, Saint John, NB, Canada. Laid up. 07.1961: Registered at Saint John as **IRVING TAMARACK**. 1969: Sold to Canadian shipbreakers and broken up. Registry closed.

(Note: Registration of this vessel at Hull as a British ship was never cancelled on sale to mercantile. It appears that certificates were not passed between respective Government departments and in consequence it was necessary to trace and rectify before registration in Canada could be effected. This resulted in a notional Bill of Sale to Henry George Pounds dated 01.09.1960 and a Bill of Sale to J D Irving Ltd dated 12.12.1958 being registered 17 07.1961. Hull registry was then closed on this date.)

EARNEST W143/A209 Assurance class tug	1270 1943 24.03.1943	03.07.1943 14.10.1943	597	142.6 33.3 11.0	C D Holmes 1350ihp 3-cyl 13.0 knots	The Admiralty, Whitehall, London

Ordered as EARNEST but name changed before completion. 08.10.1943: Registered at Hull as **EARNER**. Official No.169327. 18.10.1943: Commissioned as a rescue tug (1-3", AA weapons) (P.No.W143). 11.1943: Based Reykjavik. 31.12.1943: Sailed Reykjavik with HMS ELM (P.No.T105) (Lt K A Grant, RNR) to assist British steamer EMPIRE HOUSMAN (7359grt/1943) (Capt David J Lewis) torpedoed by U-boat (U545) but still afloat. 03.01.1944: EMPIRE HOUSMAN again torpedoed by U-boat (U744). 05.01.1944: Vessel foundered. With HMS ELM, picked up forty-five survivors and landed them at Reykjavik. 05.1944: Based Clyde. 11.06.1944: Delivered USS DONNELL (P.No.DE56) to Londonderry from Oban. 04.10.1944: In Atlantic some 800 miles west of Ireland assisted in tow of HMCS CHEBOQUE (P.No.K317) (Ty/Act/Lt Cdr M F Oliver RCNR) damaged by torpedo from U-boat (U1227). 11.10.1944: In Swansea Bay lost tow in severe weather and warship drove ashore. 12.10.1944: Refloated and delivered Port Talbot. 02.11.1944: Delivered HMS WHITAKER (P.No.K580) to Belfast for repair after damage by Hedgehog mortar bombs exploding on launcher. 1947: Decommissioned. Recommissioned manned by Royal Fleet Auxiliary (P.No.A209). 1950s: Based Rosyth. 08.12.1959: Attended North Carr Lightship which in severe weather had parted her moorings. Stood by whilst lightship anchored. 11.12.1959: Connected to lightship and delivered Leith for repair. 06.1963: Laid up at Rosyth. 12.1964: For sale. 18.01.1965: Sold to Tsavliris (Towage & Salvage) Ltd, Piræus, for £17,300. 20.01.1965: Hull registry closed. Registered at Piræus as **NISOS RODOS**. 1972: Sold to Greek shipbreakers and broken up. 1972: Registry closed.

*The Assurance class tug **Antic** (1264) photographed on the day after completion and already flying the Dutch flag.*

(Cochrane archive)

*The **Empire Sara** (1267) is seen in commercial service as **Presto** of Ellerman's Wilson Line. She is assisting the same company's **Bassano** (4986grt/46) from the River Humber into the lock at Hull.*

(Author's collection)

GILLSTONE	1271		545disp	150.0	Amos & Smith	The Admiralty,
T355	21.10.1942	19.07.1943		27.6	850ihp 3-cyl	Whitehall,
Isles class trawler	24.04.1943	13.11.1943		10.5	12.0 knots	London

Commissioned as a minesweeper (1-12pdr, AA weapons, DC) (P.No.T355). 03.1946: Sold to F Daems, Antwerp. 06.1946: Sold to AS Bergens Fiskeriseelskap (Leif Gran Kahrs), Bergen, for conversion to trawler but never completed. 10.1947: Sold to E N Kristensen, Arendal. 15.10.1947: Drifted ashore on Raunen, Jæren, en route from Bergen to Risør for rebuilding, one crewman died. Bought back by Leif Gran Kahrs, Bergen. 1948: Laid up at Fjeldberg Bruk, Stavanger. 01.1949: Sold to DS AS Anglo (Valdemar Skogland), Haugesund, but still laid up in damaged condition at Fjeldberg Bruk. 06.1952: Delivered to owner after being lengthened and converted to a refrigerated vessel by Br Lothe AS Flytedokken, Haugesund. New engine fitted - 6-cylinder Mirrlees of 810bhp. Renamed **ARGO**. Remeasured to 459grt, 144net. 1962: Sold to Freedom Line Inc, Panama. Haugesund registry closed. Registered at Panama as **FREEDOM FIRST**. 1964: Sold to Hudson Shipping Corp, Panama. Registered at Panama as **GLENROCK**. 1969: Sold to Seaway Lines Inc, Panama. Registered at Panama as **SEA ENTERPRISE**. 1970: Sold to Linea Panama Imperial Reefers S.A, Panama. Registered at Panama as **ALMIRANTE**. 04.1974: Scuttled by Dade Sports Commission south of Elliot Key, Florida, as a recreational dive site at a depth of 135ft in position 25.25N 80.07W. 1992: Damaged by Hurricane Andrew.

GRAIN	1272		545disp	150.0	Amos & Smith	The Admiralty,
T360	21.10.1942	17.08.1943		27.6	850ihp 3-cyl	Whitehall,
Isles class trawler	23.04.1943	06.12.1943		10.5	12.0 knots	London

Commissioned as a minesweeper (1-12pdr, AA weapons, DC) (P.No.T360). 16.03.1946: Sold to Italian Navy (Marina Militare) (P.No.RD309). 1965: Expended as a target.

EMPIRE HUMPHREY	1273		274	105.2	Amos & Smith	Ministry of War Transport,
169344	1943	02.09.1943		26.6	825ihp 3-cyl	London
Hoedic class tug	1943	28.12.1943		12.2	11.0 knots	

13.12.1943: Registered at Hull. Employed on coastal towage. 06.1944: Assigned to Operation Neptune - Normandy landings; SNO Corncob. 06.1944: Sailed Oban escorting Corncob (III) blockships for Normandy beaches. Towage/escort in support of Normandy landings, then coastal towage and harbour duty at Heysham. 30.08.1945: Transferred to the Admiralty and delivered by Townsend Bros, London, to Bombay for SE Asia Command. 12.1945: Delivered to Batavia; Dutch manned on harbour duties. 03.07.1946: Sold to Netherlands East Indies Government, Batavia. 28.08.1947: Hull registry closed. 08.1947: Registered at The Hague as **SUUS**. 1950: Sold to Netherland Indonesie East Indies Steenkolen Handel Maats, Amsterdam. The Hague registry closed. Registered at Amsterdam. (remained in Indonesia). 1957: Taken over by the Indonesian Government, Djakarta. 01.1961: Transferred to Perusahaan Negara Tundabara, Djakarta, Indonesia. Reported operating as **LAUT SAWU**, Indonesian flag. 1964: Transferred to Surabaja Port Authority, Surabaja, East Java. 10.2005: Removed from Lloyd's Register of Shipping - "Vessel's continued existence in doubt".

EMPIRE VINCENT	1274		275	105.3	Amos & Smith	Ministry of War Transport,
169353	1943	03.09.1943		26.6	825ihp 3-cyl	London
Hoedic class tug	1943	20.01.1944		12.2	11.0 knots	

14.01.1944: Registered at Hull. Employed on coastal towage. 06.1944: Assigned to Operation Neptune - Normandy landings. Allocated SNO Selsey. 06.06.1944: Assigned to escort Corncob (III) blockships. 08.06.1944: In Solent handling ammunition barges damaged propeller. 13.06.1944: Arrived Mulberry A towing Whale unit S16/L1. 04.06.1945: Released. 07.1945: Handed over to Pedder & Mylcreest, London, for delivery from Gibraltar to Bombay then later to Townsend Bros. Ltd, London, for delivery to Bangkok. 12.1945: Transferred to the Government of Thailand, Bangkok. 1951: Sold to the Government of Thailand. 19.04.1951: Hull registry closed. Registered at Bangkok. Later transferred to Royal Thai Navy and renamed HTMS **SAMAESAN** (P.No.YTB7). 1982: Laid up in Chao Phraya river, Bangkok. 1983: Sold to shipbreakers and broken up. Registry closed.

SESAME	1275		597	142.6	C D Holmes	The Admiralty,
W144	1943	01.10.1943		33.3	1350ihp 3-cyl	Whitehall,
Assurance class tug	09.06.1943	13.01.1944		11.0	13.0 knots	London

Commissioned as a Rescue tug (1-3", AA weapons) (P.No.W144). 06.1944: Assigned to Operation Neptune - Normandy landings. Allocated SNO Selsey. 11.06.1944: Towing Whale units to Mulberry A & B. Torpedoed by German E.boat off Normandy beaches, hit amidships on starboard side and foundered in position 49.47N 00.30W. Eighteen crew lost. Survivors picked up by HMRT STORMKING (W87) (see yard no.1258).

LONGA	1276		545disp	150.0	C D Holmes	The Admiralty,
T366	13.12.1942	15.10.1943		27.6	850ihp 3-cyl	Whitehall,
Isles class trawler	05.07.1943	11.02.1944		10.5	12.0 knots	London

Commissioned as a minesweeper (1-12pdr, AA weapons, DC) (P.No.T366). 03.1946: Transferred to the Secretary of State for Scotland (Department of Agriculture & Fisheries for Scotland) and fitted out for fishery patrol duties (unregistered). 462grt 168net. 11.1973: Sold to Thos W Ward Ltd, Sheffield, and broken up at Inverkeithing.

ORONSAY	1277		545disp	150.0	Amos & Smith	The Admiralty,
T375	13.12.1942	30.10.1943		27.6	850ihp 3-cyl	Whitehall,
Isles class trawler	17.08.1943	10.03.1944		10.5	12.0 knots	London

Commissioned as a minesweeper (1-12pdr, AA weapons, DC) (P.No.T375). 03.1946: Sold to Cie Marocaine des Pêcheries, Casablanca, Morocco, and fitted out for trawling. 10.1948: Converted to motor and fitted with 7-cylinder oil engine by British Auxiliaries Ltd, Govan. Registered at Casablanca as **MABROUK**. 439grt, 196net. 12.09.1958: Off coast of southern France started to take in water and foundered near Nice. 1958: Casablanca registry closed.

VATERSAY	1278		545disp	150.0	Amos & Smith	The Admiralty,
T375	13.12.1942	13.11.1943		27.6	850ihp 3-cyl	Whitehall,
Isles class trawler	17.08.1943	30.03.1944		10.5	12.0 knots	London

Commissioned as a minesweeper (1-12pdr, AA weapons, DC) (P.No.T378). 03.1946: Sold to Soc Nav de l'Ouest Africain, Dakar, Senegal. Fitted out for general cargo, converted to motor and fitted with 8-cylinder oil engine by Crossley Bros Ltd, Manchester. Registered at Dakar as **VOURI**. 485grt, 300net. 1954: Sold to to Huynh-Van-Gia, Saigon, Vietnam. Dakar registry closed. Registered at Saigon as **NAM VIET**. 1964: Sold to Chau Nhut Thanh, Saigon. 1972: Sold to Ho Van Tu, Saigon/Ho Chi Minh City. Converted to a tanker. 1987: Sold to Government of the Socialist Republic of Vietnam, Ho Chi Minh City. 1991: Removed from Lloyd's Register of Shipping - "Vessel's continued existence in doubt".

EMPIRE SILAS	1279		274	105.3	Amos & Smith	Ministry of War Transport,
180236	13.12.1942	13.12.1943		26.6	825ihp 3-cyl	London
Hoedic class tug	17.08.1943	20.04.1944		12.2	11.0 knots	

29.03.1944: Registered at Hull. Employed on coastal towage. 06.1944: Assigned to Operation Neptune - Normandy landings. Allocated SNO Selsey. Coastal towage in preparation for assault. 08.06.1944: Arrived off Normandy beaches. 09.06.1944: Towed ammunition barges from Solent to Juno beach. 12.1944: Harbour tug at Rouen. 04.1945: Employed on coastal towage. 20.03.1946: Transferred to Ministry of Transport, London. 03.06.1946: Sold to Fairplay Towage & Shipping Co Ltd, London for £24,000. Registered at Hull as **FAIRPLAY TWO**. 22.06.1947: Capsized in Flushing Roads, Holland, after being girded whilst towing Finnish steamer KUURTANES (3,088grt/1906) from Cardiff towards Antwerp. 1947: Salved and towed to Antwerp and declared a constructive total loss. Taken over by the Salvage Association. Repaired by Union de Remorquage et de Sauvetage S.A., Antwerp. 01.1948: Sold to Soc Cherifienne de Remorquage et d'Assistance, Casablanca, Morocco. 22.01.1948: Hull registry closed. 01.1948: Registered at Casablanca as **IFRANE**. 09.1978: Sold to Maroc Metaux, Casablanca, Morocco, for breaking up. 25.09.1978: Breaking up commenced. 1978: Registry closed.

EMPIRE BETSY	1280		274	105.3	Amos & Smith	Ministry of War Transport,
180248	1943	14.12.1943		26.6	825ihp 3-cyl	London
Hoedic class tug	1943	04.05.1944		12.2	11.0 knots	

24.03.1944: Registered at Hull. Employed on coastal towage. 06.1944: Assigned to Operation Neptune - Normandy landings. Under COTUG control. 06.06.1944: Sailed Portland escorting Corncob blockships, picked up disabled barges in Channel. 07.06.1944: Arrived Mulberry B. 06.1944: Employed off Normandy beaches. 07.1944: Employed on coastal towage. 12.09.1944: Transferred to the Admiralty. Employed on coastal towage. 04.1946: Transferred to Ministry of Transport, London. 12.1946: Sold to N V de Bataafsche Petroleum Maats, The Hague, Holland. 09.01.1947: Hull registry closed. Registered at The Hague as **SOEGIO**. 12.02.1948: Mined in Macassar Strait, Borneo, and foundered in position 02.36S 116.33E. 1948: Registry closed.

ENVOY	1281		868disp	160.0	C D Holmes	The Admiralty
W165/A165	1943	11.02.1944		36.0	1700ihp 3-cyl	Whitehall
Envoy class tug	02.10.1943	17.05.1944		15.0	13.0 knots	London

Commissioned as a Rescue tug (1-3", AA weapons) (P.No.W165). 06.1944: Assigned to Operation Neptune - Normandy landings. Allocated SNO Selsey. 10.06.1944: Arrived Mulberry A with Whale tow. 24.06.1944: With Dutch motor tug THAMES (602grt/1938) towed light cruiser HMS SCYLLA (P.No.98) damaged by mine, disabled without electrical power and delivered to the Solent. 1946: Paid off and placed in reserve. 1947: Registered at Harwich. Official No.163014; 762grt, 78net. 05.1948: Surveyed and classified at London. 1949: Manned by Royal Fleet Auxiliary (P.No.A165). 1965: Sold to Loucas Matsas & Sons, Piræus, Greece. Harwich registry closed. Registered at Piræus as **MATSAS**. 1968: Registered at Piræus as **GEORGIOS L MATSAS**. 1973: Sold to Oldscrap Lendaris Carvoudakis, Greece. 27.07.1973: Arrived Perama for breaking up. 1973: Registry closed.

ENTICER	1282		868disp	160.0	C D Holmes	The Admiralty
W166	1943	11.03.1944		36.0	1700ihp 3-cyl	Whitehall
Envoy class tug	20.10.1943	27.06.1944		15.0	13.0 knots	London

Commissioned as a Rescue tug (1-3", AA weapons) (P.No.W165). 1946: Based Hong Kong manned by Yard Craft officers (Master Albert G Adcock), local crew. 12.1946: Sailed Hong Kong in company with frigate HMS ALACRITY (P.No.U60) to go to the aid of steamer ROSEBANK (2410grt/1920) disabled with damaged rudder in very heavy weather off Hainan Island, South China Sea. 21.12.1946: Off east coast of Hainan Island, tug shipped several heavy seas and subsequently foundered. ALACRITY picked up twelve out of crew of twenty-two but only one survived. (ALACRITY connected to ROSEBANK but could only hold her head to wind until cruiser HMS EURYALUS (P.No.42) diverted to scene, picked up tow and delivered to Hong Kong).

HELLISAY	1283		545disp	150.0	Amos & Smith	The Admiralty,
T391	16.04.1943	27.03.1944		27.6	850ihp 3-cyl	Whitehall,
Isles class trawler	17.11.1943	14.07.1944		10.5	12.0 knots	London

Commissioned as a minesweeper/danlayer (AA weapons) (P.No.T391). 06.1947: Sold to A G Chalaris & Co, Piræus, Greece. Lengthened to 153.8ft; 443grt, 182net. Converted to general cargo. Registered at Piræus as **ELPIS**. 1954: Sold to P, A & G Dacoutros, Valetta/Sliema, Malta. Registered at Malta as **ELPIS II**. Official No.191783. Remeasured to 503grt, 197net. 1957: Removed from Lloyd's Register of Shipping.

HERMETRAY T392 Isles class trawler	1284 16.04.1943 17.11.1943	11.04.1944 17.08.1944	545disp	150.0 27.6 10.5	Amos & Smith 850ihp 3-cyl 12.0 knots	The Admiralty, Whitehall, London

Commissioned as a danlayer (AA weapons) (P.No.T392).   25.04.1947: Sold to N V Scheeppvaart Maats Andomeda, IJmuiden. Converted to a steam trawler. Lengthened to 153.8ft; 453grt, 160net. Registered at IJmuiden as **COIMBRA**.   1948: Sold to N V Vissch Maats "Prinses Beatrix", IJmuiden.   24.12.1952: Sold to Lloyd Seeschiffahrt A G, Basle, Switzerland.   12.1952: IJmuiden registry closed.
02.01.1953: Registered at Basle as **FURKA**. Fitted out for general cargo by Mützelfeldtwerft GmbH, Cuxhaven; converted to motor and fitted with 6-cylinder oil engine by by Swiss Loco & Mach Wks (Sulzer), Winterthur. Lengthened to 178.4ft, now 677grt, 377net.
15.10.1954: Sold to Lloyd Seeschiffahrt GmbH, Hamburg.   14.06.1957: Sold to Edmond H Smith, La Habana, Cuba. Basle registry closed.
Registered at Panama as **NENTER**.   1964: Sold to Motonaves del Golfo S A, Panama.   1965: Sold to Navieros de Yucatan S.A, Progreso, Mexico. Panama registry closed. Registered at Progreso as **T'HO**.   16.03.1976: Stranded on Colson Cay whilst on passage from Agua Dulce towards Belize City, making water throughout and abandoned. Declared a total loss. Set on fire. 1976: Registry closed.

EMPIRE CHRISTOPHER 180291 Hoedic class tug	1285 1943 1943	09.05.1944 30.08.1944	275	105.3 26.6 12.2	Amos & Smith 825ihp 3-cyl 11.0 knots	Ministry of War Transport, London

01.08.1944: Registered at Hull.   10.1944: Delivered to Pedder & Mylchreest Ltd, London, for delivery to Rangoon. Employed on miscellaneous naval duties in Burma and Singapore.   21.04.1946: Mined off Maungmagan Bay, Gulf of Martaban, Burma, whilst on passage from Rangoon towards Singapore and foundered quickly in position 14.09N 98.03E. Seventeen lost out of crew of twenty-six.
26.09.1946: Registry closed - "...vessel sunk by mine".

EMPIRE JOSEPHINE 180297 Hoedic class tug	1286 1943 1943	09.05.1944 15.09.1944	274	105.3 26.6 12.2	Amos & Smith 825ihp 3-cyl 11.0 knots	Ministry of War Transport, London

22.08.1944: Registered at Hull.   09.1944: Delivered to Pedder & Mylchreest Ltd, London, for delivery to Sydney, Australia, and employed on Admiralty service with Pacific fleet.   09.1945: Arrived Hong Kong. Employed as dockyard tug.   1946: Released from Admiralty service and transferred to Ministry of Transport.   01.05.1946: Sold to the Hong Kong Government, Hong Kong, for use by Hong Kong Police.
29.10.1946: Hull registry closed.   1965: Sold to Yau Wing Co Ltd, Hong Kong. Registered at Hong Kong as **YAU WING No. 25**.
1966: Sold to Cheung Chau Shipping & Trading Co, Hong Kong.   1966: Sold to Pan Asiatic Lines Inc of Carolus SA, Panama. Hong Kong registry closed. Registered at Panama as **FEDREDGE JOSEPHINE**.   1967: Sold to Wo Hing Co, Hong Kong, for breaking up.
09.1967: Commenced breaking up.   1967: Registry closed.

ENIGMA W175 Envoy class tug	1287 06.04.1943 12.02.1944	11.03.1944 19.10.1944	868disp	160.0 36.0 15.0	C D Holmes 1700ihp 3-cyl 13.0 knots	The Admiralty Whitehall London

Commissioned as a rescue tug (1-3", AA weapons) (P.No.W175).   1947: At Singapore employed as fleet target towing and rescue tug.
13.10.1952: Refitted for harbour work.   26.07.1962: Commissioned with crew ex HMS NIMBLE (890disp/1941) and later sailed for Devonport.   08.10.1962: Arrived Devonport and paid off.   1964: Sold to Nicholas E Vernicos Shipping Co Ltd, Piræus, Greece. Registered at Piræus as **VERNICOS**. 793grt; 1620ihp.   1972: Sold to shipbreakers in Greece and broken up.   1972: Registry closed.

ENFORCER W177 Envoy class tug	1288 06.04.1943 13.03.1944	22.07.1944 22.11.1944	868disp	160.0 34.5 15.0	C D Holmes 1700ihp 3-cyl 13.0 knots	The Admiralty Whitehall London

Commissioned as a rescue tug (1-3", AA weapons) (P.No.W177).   06.1947: Registered at Harwich. Official No.163015. 762grt, 78net.
Allocated to CinC Portsmouth.   04.1949: Surveyed at London. Manned by Royal Fleet Auxiliary.   26.01.1950: Transferred to Rosyth for seagoing duties.   1963: Sold to BISCO and allocated to James A White & Co Ltd, St Davids, for breaking up.   1963: Registry closed.

IMERSAY J422 Isles class trawler	1289 16.04.1943 12.04.1944	21.08.1944 05.12.1944	545disp	150.0 27.6 10.5	Amos & Smith 850ihp 3-cyl 12.0 knots	The Admiralty, Whitehall, London

Commissioned as a danlayer (AA weapons) (P.No.J422). Attached to M/S Flotillas East Indies Fleet.   1946: Remained in Far East on mine clearance.   1958: At Malta and offered for sale.   01.1959: Sold to John S Latsis, Athens & London. Fitted out for general cargo. Registered at Piræus as **MARIA**; 452grt.   1964: Sold to Piangos Bros, Piræus, Greece. Registered at Piræus as **MICHAEL**; 488grt.   1965: Converted to motor and fitted with 6-cylinder 600bhp oil engine by MWM, Mannheim.   27.06.1969: Damaged by fire.   1969: Sold to Ali Riza Yakup Aksoy, Gallipoli, Turkey. Repaired and returned to service. Piræus registry closed. Registered at Gallipoli as **MEHMET AKSOY**.
1974: Sold to Iltas Kollektif Sirketi, Istanbul. Registered at Istanbul.   1983: Sold to Ali Riza Yakup Aksoy, Gallipoli.   1992: Remeasured to 495grt, 275net, 850dwt.   1993: Sold to Mehmet Aksoy, Istanbul. Registered at Istanbul as **AKSU-1**.   1995: Sold to Huseyin Aksu, Istanbul.
2002: Removed from Lloyd's Register of Shipping - "Vessel's continued existence in doubt".

LINGAY J423 Isles class trawler	1290 16.04.1943 12.04.1944	06.09.1944 29.12.1944	545disp	150.0 27.6 10.5	Amos & Smith 850ihp 3-cyl 12.0 knots	The Admiralty, Whitehall, London

Commissioned as a danlayer (AA weapons) (P.No.J423). Attached to M/S Flotillas East Indies Fleet.   1946: Remained in Far East on mine clearance.   1947: Sold to British Wheeler Process Ltd, Liverpool.   1948: Converted to a tank cleaning vessel. Registered at Liverpool as **TULIPDALE**. Official No.188416; 459grt, 170net.   1965: Sold to Belgian shipbreakers.   24.09.1965: Arrived Antwerp under tow for breaking up.   09.1965: Registry closed.

EMPIRE JENNY 180388 Hoedic class tug	1291 1944 1944	04.10.1944 16.01.1945	274	105.3 26.6 12.2	Amos & Smith 825ihp 3-cyl 11.0 knots	Ministry of War Transport, London

07.12.1944: Registered at Hull. 01.1945: Delivered to Pedder & Mylchreest Ltd, London, for delivery to Trincomalee.
31.05.1946: Transferred to the Admiralty. Yard Craft master, locally entered crew. 03.06.1947: Hull registry closed. 29.08.1947: Renamed **AID** (unregistered). 1952: Transferred to Malta and placed in reserve. 08.1958: Transferred to Portsmouth. 26.07.1960: Sold to Henry George Pounds, Paulsgrove, Portsmouth. 1961: Sold to J D Irving Ltd, Saint John, NB, Canada. Registered at Saint John as **IRVING TEAK**. 1977: Sold to shipbreakers in Canada. 06.06.1977: Commenced breaking up. 1977: Registry closed.

EMPIRE BARBARA 180320 Hoedic class tug	1292 1944 1944	05.10.1944 06.02.1945	275	105.3 26.6 12.2	Amos & Smith 825ihp 3-cyl 11.0 knots	Ministry of War Transport, London

20.04.1944: Registered at Hull. 01.1945: Delivered to Pedder & Mylchreest Ltd, London, for delivery to Trincomalee. Allocated to CinC East Indies Station. Locally entered crew. 31.03.1947: Transferred to the Admiralty. 03.06.1947: Hull registry closed.
29.08.1947: Renamed **ADEPT**. 03.12.1957: Sold to the Government of Ceylon, Colombo, and allocated to the Royal Ceylon Navy. Renamed **ALIYA**. 1964: Refitted for further service. 1978: Sold to Steel Corporation of Sri Lanka and broken up.

ENCHANTER W178 Envoy class tug	1293 06.04.1943 06.1944	02.11.1944 27.03.1945	868disp	160.0 36.0 15.0	C D Holmes 1700ihp 3-cyl 13.0 knots	The Admiralty Whitehall London

06.04.1945: Commissioned as a Rescue tug (1-3", AA weapons) (P.No.W178). 27.06.1947: Sold to United Towing Co Ltd, Hull.
06.1947: Registered at Hull as **ENGLISHMAN**. Official No.181307; 762grt, 68net. 28.03.1950: Remeasured to 716grt 41net.
29.03.1962: Sold to Suprema Cia Naviera S A, Panama. 03.1962: Hull registry closed. Registered at Panama as **CINTRA**. 1968: Sold to Tsavliris (Salvage & Towage) Ltd, Piræus. Panama registry closed. Registered at Piræus as **NISOS SKIATHOS**. 09.1972: Sold to I Kahmertzoglou and broken up at Perama. 1972: Registry closed.

ENCORE W179/A379 Envoy class tug	1294 06.04.1943 24.07.1944	02.12.1944 30.04.1945	868disp	160.0 36.0 15.0	C D Holmes 1700ihp 3-cyl 13.0 knots	The Admiralty Whitehall London

11.05.1945: Commissioned as a Rescue tug (1-3", AA weapons) (P.No.W179). 1959: Registered at London. Official No.301071; 780grt. Manned by Royal Fleet Auxiliary (P.No.A379). Based Hong Kong. 01.1962: Sailed Hong Kong following Hong Kong tug TAI KOO (812grt/1950) to go to the assistance of STANVAC SUMATRA (10202grt/1949), broken in two on 27.01.1962 in very heavy weather some 300 miles south-east of Saigon. 28.01.1962: Crew taken off after part by USS COOK (APD-2130). Located in position 09.5N 109.55E. After part taken in tow by TAI KOO. 07.02.1962: Delivered Singapore Roads. ENCORE diverted to Singapore to fuel and then picked up fore part. 12.02.1962: Delivered Singapore Roads. 06.11.1967: Sold to Selco (Singapore) Ltd, Singapore. London registry closed. Registered at Singapore as **SALVALIANT**. 1969: Laid up. 1972: Sold to Hong Huat (Pte) Ltd, Singapore. 08.09.1972: Breaking up commenced at Jurong. 1972: Registry closed.

ORSAY J450 Isles class trawler	1295 16.04.1943 07.09.1944	01.01.1945 31.05.1945	545disp	150.0 27.6 10.5	Amos & Smith 850ihp 3-cyl 12.0 knots	The Admiralty, Whitehall, London

Commissioned as a danlayer (AA weapons) (P.No.J450). 1946: Designated as a minesweeper (P.No.M341). 03.09.1957: Sold to Cossira S.p.A Di Navigazione, Portoferraio, Italy. Registered as **ANTILOPE**. 1971: Converted to motor and fitted with 12-cylinder 900bhp oil engine by General Motors Corp, La Grange, Illinois, USA. 21.01.1975: Sold to C N Della Palmaria, La Spezia. 20.02.1975: Breaking up commenced. Registry closed.

RONAY J429 Isles class trawler	1296 16.04.1943 07.09.1944	15.02.1945 08.06.1945	545disp	150.0 27.6 10.5	Amos & Smith 850ihp 3-cyl 12.0 knots	The Admiralty, Whitehall, London

Commissioned as a danlayer (AA weapons) (P.No.J429). 12.03.1967: Sold to West of Scotland Shipbreaking Co Ltd, Troon, for breaking up. 13.04.1967: Arrived Troon.

EMPIRE FLORA 180438 Stella class tug	1297 1944 1944	16.03.1945 24.07.1945	292	116.0 28.0 12.7	Franklin USA 525ihp 3-cyl 11.0 knots	Ministry of War Transport, London

Yard Nos.1297 to 1304: Triple expansion engines made by Franklin Machine and Foundry Corporation, Providence, Rhode Island, USA (Under Lease Lend Scheme surplus to US requirements). All this class built at Selby originally intended for Far East deployment.
25.06.1945: Registered at Hull. General towage duties. 04.1946: Transferred to Ministry of Transport, London. 16.03.1948: Sold to Risdon Beazely Ltd, Southampton. 28.05.1948: Registered at Hull as **TOPMAST 14**. 05.03.1949: Sold to Soc Anon Rimorchiatori Riuniti "Panfido & Co", Venice, Italy. 05.03.1949: Hull registry closed. Registered at Venice as **TAURUS**. 1976: Owners re-styled Panfido e Cia, Societa per Azioni Rimorchiatori Riuniti, Venice. 08.1984: Sold to shipbreakers in Porto Nogaro and broken up by Eurofer S.P.A. 1984: Registry closed.

Longa (1276), an Isles class trawler.

(Cochrane archive)

Envoy (1281) which gave its name to a class of tugs. A barrage balloon is visible in the background.

(Cochrane archive)

Orsay (1295), another Isles class trawler as a danlayer and seen from the starboard side.

(Cochrane archive)

EMPIRE STELLA 180448	1298 1944		325	116.0	Franklin USA	Ministry of War Transport,
		16.03.1945		28.0	525ihp 3-cyl	London
Stella class tug	1944	27.08.1945		12.7	11.0 knots	

03.08.1945: Registered at Hull. General towage duties. 01.01.1946: Suffered serious damage following a boiler explosion in the Thames Estuary when oil was present in feed water; four crew killed, two missing and others injured. Survivors including master, Capt Woolnough, taken off by Belgian trawler MARIE LOUISE (P.No.Z238) (258grt/1918). 04.1946: Transferred to Ministry of Transport, London.
01.05.1946: Towed to Hull and laid up. 15.05.1946: Sold to United Towing Co Ltd, Hull, for £15,000. Registered at Hull as **SERVICEMAN**.
09.1946: Completed repairs and re-engined with 750ihp triple expansion engine by Alex Hall & Co Ltd, Aberdeen (originally to have been fitted in EMPIRE KEITH). 1961: Converted to motor and fitted with 6-cylinder 1575bhp oil engine by British Polar Engines Ltd, Glasgow. 13 knots. 330grt, 55net. 11.1969: Sold to Rimorchiatori Sardi Spa, Cagliari, Sardinia. 27.11.1969: Hull registry closed. Registered at Cagliari as **POETTO** (IMO 5320895). 2015: Still in service.

EMPIRE SHEILA 180451	1299 1944		292	116.0	Franklin USA	Ministry of War Transport,
		30.03.1945		28.0	525ihp 3-cyl	London
Stella class tug	1944	13.09.1945		12.7	11.0 knots	

06.09.1945: Registered at Hull. 12.09.1945 - 05.1946: Employed on coastal towage. 04.1946: Transferred to Ministry of Transport, London. 22.10.1947: Sold to Overseas Towage & Salvage Ltd, London. 14.03.1949: Hull registry closed. Registered at London as **SHEILIA**. 1950: Sold to Société Chérifienne de Remorquage et d'Assistance, Casablanca, Morocco. London registry closed. Registered at Casablanca as **SIDI BELYOUT**. 1956: Converted to motor and fitted with 8-cylinder 1575bhp oil engine by Motorenwerke Mannheim A G (MWM), Mannheim. 311grt, 131net. 1961: Sold to Rimorchiatori Sardi SpA, Cagliari, Sardinia. Casablanca registry closed. Registered at Cagliari as **TIRSO**. 1997: Removed from Lloyd's Register of Shipping - "Vessel's continued existence in doubt".

EMPIRE CLARA 180453	1300 1944		292	116.0	Franklin USA	Ministry of War Transport,
		28.04.1945		28.0	525ihp 3-cyl	London
Stella class tug	1944	05.10.1945		12.7	11.0 knots	

11.09.1945: Registered at Hull. General towage duties. 04.1946: Transferred to Ministry of Transport, London. 28.07.1947: Sold to United Towing Co Ltd, Hull, for £8,000. 13.12.1947: Registered at Hull as **AIRMAN**. 1949: Re-engined with 660ihp triple expansion engine by Amos & Smith Ltd, Hull. 11.5 knots. 23.05.1950: Re-registered following alteration of particulars. 333grt. 1967: Sold to Hughes Bolckow Ltd, Blyth, in en bloc sale with RIFLEMAN (see yard no.1302) and GUARDSMAN (see yard no.1304). 18.11.1967: Arrived Blyth under tow for breaking up. 29.01.1968: Registry closed - "Vessel broken up".

EMPIRE MARTHA 180459	1301 1944		292	116.0	Franklin USA	Ministry of War Transport,
		28.04.1945		28.0	525ihp 3-cyl	London
Stella class tug	1944	07.11.1945		12.7	11.0 knots	

02.11.1945: Registered at Hull. General towage duties. 04.1946: Transferred to Ministry of Transport, London. 25.02.1947: Sold to James Contracting & Shipping Co Ltd, London. 25.02.1947: Hull registry closed. Registered at London as **FOREMOST 106**. 1949: Sold to Remorquage Letzer Soc Anon, Antwerp, Belgium. London registry closed. Registered at Antwerp as **GEORGES LETZER**.
1964: Converted to motor fitted with 8-cylinder 4000bhp oil engine by Klöckner-Humbolt-Deutz, Cologne. 14 knots. 1977: Owners re-styled Union de Remorquage et de Sauvetage SA, Antwerp. 1981: Rebuilt at Rupelmonde, Belgium, and fitted with Kort nozzle. 1992: Sold to Northern Europe Shipping Ltd, Antwerp. Registered at Antwerp as **HILDE**. 1994: Sold to Jacques Bakker & Zonen, Bruges. 06.1994: Arrived Bruges for breaking up. 06.1994: Registry closed.

EMPIRE VERA 180466	1302 1944		292	116.0	Franklin USA	Ministry of War Transport,
		14.05.1945		28.0	525ihp 3-cyl	London
Stella class tug	1945	11.12.1945		12.7	11.0 knots	

06.12.1945: Registered at Hull. General towage duties. 04.1946: Transferred to Ministry of Transport, London. 28.07.1947: Sold to United Towing Co Ltd, Hull. 10.10.1947: Registered at Hull as **RIFLEMAN**. 1949: Re-engined 660ihp triple expansion engine by Amos & Smith Ltd, Hull. 11.5 knots. 20.04.1950: Re-registered following alteration of particulars. 333grt. 1967: Sold to Hughes Bolckow Ltd, Blyth, in en bloc sale with AIRMAN (see yard no.1300) and GUARDSMAN (see yard no.1304). 18.11.1967: Arrived under tow at Blyth for breaking up. 28.01.1968: Registry closed - "Vessel broken up".

EMPIRE GRETA 180468	1303 1945		292	116.0	Franklin USA	Ministry of War Transport,
		12.06.1945		28.0	525ihp 3-cyl	London
Stella class tug	1945	29.01.1946		12.7	11.0 knots	

10.01.1946: Registered at Hull. 04.1946: Transferred to Ministry of Transport, London. General towage duties. 04.02.1947: Sold to James Contracting & Shipping Co Ltd, London. 18.02.1947: Hull registry closed. 02.1947: Registered at London as **FOREMOST 105**. 1948: Sold to Cork Harbour Commissioners, Cork, for £25,000. 1949: London registry closed. Registered at Cork as **FRANCIS HALLINAN**. 11.1968: Sold to L E Adams, Okehampton, Devon. 12.1968: On passage Cork towards Penzance disabled with flooding in engine and boiler room. Dutch salvage tug GRONINGEN (598grt/1963) connected and delivered to Penzance. 12.1968: Sailed Penzance for Falmouth after repair. 01.1969: Arrested by the Admiralty Marshal on behalf of GRONINGEN owners for non payment of salvage dues. 12.02.1969: In stormy weather broke away from mooring and adrift in Falmouth harbour. Recovered and secured by local tugs. 1971: Transferred to Sapsin Property Co Ltd, Okehampton, Devon. 1972: Sold to Flamecap Ltd, Rochester, Kent, but resold to Hierros Varela, Bilbao, Spain, for breaking up. 07.03.1973: Sailed Falmouth for Bilbao in tow of motor tug FAIRPLAY X (298grt/1967) in tandem with ST LEVAN (160grt/1942) also for breaking up. 12.03.1973: Arrived Bilbao. 19.04.1973: Breaking completed. Registry closed.

EMPIRE NINA 180474	1304 1945		292	116.0	Franklin USA	Ministry of War Transport,
		12.06.1945		28.0	525ihp 3-cyl	London
Stella class tug	1945	02.03.1946		12.7	11.0 knots	

18.02.1946: Registered at Hull. 04.1946: Transferred to Ministry of Transport, London. General towage duties. 10.03.1947: Sold to United Towing Co Ltd, Hull for £10,500. 18.07.1947: Registered at Hull as **GUARDSMAN**. Re-engined with 750ihp triple expansion engine by Alex Hall & Co Ltd, Aberdeen (originally to have been fitted in EMPIRE GEORGE). 11.5 knots. 18.07.1967: Re-registered following alteration of particulars. 329grt. Sold to Hughes Bolckow Ltd, Blyth, in en bloc sale with AIRMAN (see yard no.1300) and RIFLEMAN (see yard no.1302). 18.11.1967: Arrived under tow at Blyth for breaking up. 29.01.1968: Registry closed - "Vessel broken up".

Empire Stella (1298) was another tug to give her name to a class. She is still in service in Sardinia.

(Cochrane archive)

Empire Nina (1304), another Stella class tug, became **Guardsman** following sale to United Towing and she is seen assisting the **Empire Star** (11,861grt/1946) in the lock at Hull. Readers will note several modifications to the original design.

(Author's collection)

Empire Mayring (1305) was a cargo vessel designed for service in the Far East.

(Cochrane archive)

EMPIRE MAYRING	1305		395	144.0	Amos & Smith	Ministry of War Transport,
180464	1945	10.08.1945		27.1	375ihp 3-cyl	London
Cargo vessel	1945	08.01.1946		8.0	9.0 knots	

Shelterdeck 'C' type oil burning steamer designed for service in the Far East.

23.11.1945: Registered at Hull.   04.1946: Transferred to Ministry of Transport, London.   12.09.1946: Sold to Lloyds Albert Yard & Motor Packet Services Ltd, Southampton.   16.09.1946: Sold to Ta Hing Co (Hong Kong) Ltd, Hong Kong.   24.10.1946: Registered 565grt, 239net following survey at Southampton.   23.12.1946: Hull registry closed.   12.1946: Registered at Hong Kong as **SING HING**.   1951: Sold to Pakistan Steam Navigation Co Ltd, Chittagong.   Hong Kong registry closed.   Registered at Chittagong as **ISLAMABAD**.   1972: Sold to Bangladesh Steam Navigation Co Ltd, Chittagong.   08.2007: Removed from Lloyd's Register of Shipping.

EMPIRE MAYROVER	1306		394	144.0	Amos & Smith	Ministry of War Transport,
181262	1945	08.09.1945		27.1	375ihp 3-cyl	London
Cargo vessel	1945	31.05.1946		8.0	9.0 knots	

Shelterdeck 'C' type oil burning steamer designed for service in the Far East.

Completed for Ministry of Transport, London.   24.05.1946: Registered at Hull.   26.08.1946: Sold to Lloyds Albert Yard & Motor Packet Services Ltd, Southampton.   31.08.1946: Sold to Ta Hing Co (Hong Kong) Ltd, Hong Kong.   10.1946: Registered 565grt, 261net following survey at Southampton.   23.12.1946: Hull registry closed.   12.1946: Registered at Hong Kong as **WA HING**.   1949: Sold to India General Navigation & Railway Co Ltd, Calcutta. Registered at Hong Kong as **MUMTAZ**.   1949: Registered at Chittagong.   1961: Sold to Pakistan River Steamers Ltd, Chittagong.   1972: Sold to Bangladesh River Steamers Ltd, Chittagong.   1973: Sold to Government of Bangladesh, Chittagong (Bangladesh River Steamers Ltd).   Registered at Chittagong as **C5-203**.   1975: Sold to Bangladesh Inland Water Transport Corporation, Chittagong.   1992: Broken up at Chittagong.   Registry closed.

EMPIRE HELEN	1307		275	105.3	Amos & Smith	Ministry of War Transport,
180889	1945	11.07.1945		26.6	825ihp 3-cyl	London
Hoedic class tug	1945	14.06.1946		12.2	11.0 knots	

Completed for Ministry of Transport, London.   06.1946: Registered at London.   General towage duties.   1946: Sold to Overseas Towage & Salvage Co Ltd, London for £27,500.   Registered at London as **NEREIDIA**.   1951: Sold to Compagnie de Remorquage et de Sauvetage 'Les Abeilles', Le Havre, France.   London registry closed.   Registered at Le Havre as **ABEILLE No. 7**.   1961: Sold to Rimorchiatori Sardi, SpA, Cagliari, Sardinia.   Le Havre registry closed.   Registered at Cagliari as **SERGIO**.   1984: Sold to shipbreakers in Porto Torres, Sardinia, and broken up.   Registry closed.

EMPIRE SIMON	1308		275	105.3	Amos & Smith	Ministry of War Transport,
180904	1945	11.07.1945		26.6	825ihp 3-cyl	London
Hoedic class tug	1945	10.07.1946		12.2	11.0 knots	

Purchased by Overseas Towage & Salvage Co Ltd, London, for £27,500 and completed as **SIMONIA**.   Registered at London.   1951: Sold to Compagnie de Remorquage et de Sauvetage 'Les Abeilles', Le Havre, France.   London registry closed.   Registered at Le Havre as **ABEILLE No. 8**.   1966: Sold to Gaetano d'Alessio, Livorno, Italy.   Le Havre registry closed.   Registered at Livorno as **ANTONIO D'ALESIO**. 1975: Sold to Rimorchiatori Calabresi SpA, Livorno.   Registered at Livorno as **GRECALE**.   1997: Removed from Lloyd's Register of Shipping - "Vessel's continued existence in doubt".

ST BARTHOLOMEW	1309		579	177.6	Amos & Smith	Saint Andrew's Steam
180475	1945	1945	216	30.2	1000ihp 3-cyl	Fishing Co Ltd,
H216 Steam trawler	1945	21.02.1946		15.1	12.7 knots	Hull

19.02.1946: Registered at Hull (H216).   26.02.1946: Sailed on first trip (Sk Arthur J Lewis).   19.03.1946: Landed record catch, 4223 kits £17,302 gross.   17.08.1946: Sailed for Bear Island grounds.   06.09.1946: Landed in Germany for The United Nations Relief and Rehabilitation Administration (famine relief), 3717 kits, £8,365 gross.   14.09.1946: Sold to Charleson-Smith Trawlers Ltd, Hull. 15.10.1946: Registered at Hull as **STELLA ARCTURUS** (H216).   19.03.1948: Sold to East Riding Trawlers Ltd, Hull.   05.12.1950: Sold to Trawlers Grimsby Ltd, Grimsby.   10.04.1953: Sold to Derwent Trawlers Ltd, Grimsby.   09.09.1956: Disabled off Spitzbergen (Sk Don Tennyson) with engine problem; trawler BAYELLA (H72) (Sk C Drever) connected; tow parted twice but delivered to Humber in four days. 01.10.1964: Sold to Alsey Steam Fishing Co Ltd, Grimsby.   01.01.1965: Sold to Hudson Fishing Co Ltd, Hull.   06.01.1966: Registered at Hull as **ROSS ARCTURUS** (H216) (Ross Group funnel colours adopted for all 'Ross' prefixed vessels).   31.03.1966: Sold to Hudson Brothers Trawlers Ltd, Hull.   13.04.1967: Sold to Boyd Line Ltd, Hull.   05.05.1967: Registered at Hull as **ARCTIC OUTLAW** (H216). 02.07.1968: Landed and laid up for disposal.   03.10.1968: Sold to P & W MacLellan Ltd, Bo'ness, for breaking up. 19.03.1969: Hull registry closed.

ST MARK	1310		579	177.6	C D Holmes	Saint Andrew's Steam
180477	1945	06.11.1945	216	30.2	1000ihp 3-cyl	Fishing Co Ltd,
H218 Steam trawler	01.03.1945	01.03.1946		15.1	12.5 knots	Hull

28.02.1946: Registered at Hull (H218).   Insured for £97,000.   05.03.1946: Sailed Hull for Norwegian coast, first trip (Sk R G Cook). 28.03.1946: At Hull landed a near record catch, 3803 kits, £15,079.   28.10.1946: Sk Cook had completed 9 trips - 29,824 kits, £87,003. 07.12.1946: Sold to Hudson Bros Trawlers Ltd, Hull.   23.01.1947: Registered at Hull as **CAPE TRAFALGAR** (H218).   09.1949: Converted to burn oil fuel.   11.01.1955: Sold to West Dock Steam Fishing Co Ltd, Hull, for £81,000 plus £10,536 fees.   05.02.1955: Registered at Hull as **AUBURN WYKE** (H218).   05.1959: Company and four trawlers taken over by Associated Fisheries Ltd, Hull.   14.05.1959: Sold to Boyd Line Ltd, Hull, for £71,834.   28.05.1959: Registered at Hull as **ARCTIC HUNTER** (H218).   09.1968: Sold to Jos de Smedt, Antwerp, Belgium.   21.10.1968: Arrived Antwerp for breaking up.   23.10.1968: Hull registry closed.

*A fine view of **St Bartholomew** (1309) in the Humber.*

(Cochrane archive)

NORTHELLA 180482 H244 Steam trawler	1311 1945 1945	06.12.1945 05.04.1946	579 216	177.6 30.2 15.1	Amos & Smith 1000ihp 3-cyl 12.5 knots	J Marr & Son Ltd, Fleetwood

Contract price £65,125. 26.03.1946: Registered at Hull (H244). 22.05.1946: On second trip Sk W Drever landed record catch of 4319 kits, £12,842. 19.08.1946: Sailed for Bear Island grounds. 08.09.1946: Landed in Germany for The United Nations Relief and Rehabilitation Administration (famine relief), 3861 kits, £8,587. 25.06.1948: Sold to East Riding Trawlers, Hull, for £117,500. 30.08.1948: Registered at Hull as **STELLA CANOPUS** (H244). 01.1950: Converted for burning oil fuel. 05.12.1950: Sold to Trawlers Grimsby Ltd, Grimsby. 10.04.1953: Sold to Derwent Trawlers Ltd, Grimsby. 20.02.1959: Company re-styled Ross Trawlers Ltd, Grimsby. 28.03.1963: Sold to Hudson Bros Trawlers Ltd, Hull. 18.11.1964: Sold to Alsey Steam Fishing Co Ltd, Grimsby. 26.11.1965: Registered at Hull as **ROSS CANOPUS** (H244). 31.03.1966: Sold to Hudson Bros Trawlers Ltd, Hull. 1967: Sold to Scrappingco Srl, Antwerp, Belgium. 21.07.1967: Hull registry closed. 22.07.1967: Sailed Hull for Ghent under tow for breaking.

DANUBE VII 180857 Steam tug	1312 1945 1945	05.01.1946 30.04.1946	237	110.6 27.6 12.2	C D Holmes 900ihp 3-cyl 12.0 knots	Tilbury Contracting & Dredging Co Ltd, London

05.1946: Registered at London. 1965: Sold to Westminster Dredging Co Ltd, London. 1969: Sold to Soc Anon Italiana Lavoriedili Marittimi, Palermo, Sicily. London registry closed. Registered at Palermo as **GIOVE SAILEM**. 1987: Sold to Acc Ferriere, Palermo. 24.10.1987: Sold to shipbreakers in Sicily and broken up in Palermo. Registry closed.

DANUBE VIII 180863 Steam tug	1313 1945 1945	05.01.1946 15.05.1946	237	110.6 27.6 12.2	C D Holmes 900ihp 3-cyl 12.0 knots	Tilbury Contracting & Dredging Co Ltd, London

05.1946: Registered at London. 1965: Sold to Westminster Dredging Co Ltd, London. 28.09.1965: Inbound in fog off South Oaze Buoy, Thames estuary towing spoil hopper BLACK DEEP (914grt/1925) both tug and tow in collision with outbound sludge carrier SIR JOSEPH RAWLINSON which foundered rapidly with loss of nine crew. 1968: Sold to Scrappingco NV, Antwerp, Belgium. 09.1968: Arrived Boom and commenced breaking up. 09.1968: Registry closed.

EMPIRE HEDDA Improved Stella class tug	1314 1945 1945	04.02.1946 27.09.1946	327	116.0 27.6 12.7	Amos & Smith 800ihp 3-cyl 12.0 knots	Ministry of War Transport, London

Laid down as EMPIRE DOREEN. Launched as **EMPIRE HEDDA** for Ministry of Transport, London. Completed as **ATLAS** for Bergnings och Dykeri A/B 'Neptun', Stockholm, Sweden. 09.1946: Registered at Stockholm. 1965: Sold to Rimorchiatori Sardi, Cagliari, Sardinia. Stockholm registry closed. Registered at Cagliari as **MAROSA**. 1985: Sold to shipbreakers in Italy and broken up at Cagliari, Sardinia. Registry closed.

EMPIRE JUNA Improved Stella class tug	1315 1945 1945	08.02.1946 15.08.1946	327	116.0 27.6 12.7	Amos & Smith 800ihp 3-cyl 12.0 knots	Ministry of War Transport, London

Completed for Ministry of Transport, London. 10.07.1946: Registered at Hull. 11.10.1946: Sold to Government of Nigeria Ports Authority, Lagos, Nigeria. 17.10.1946: Hull registry closed. 10.1946: Registered at Lagos as **BALBUS**. 296grt. 1967: Sold to Tsavliris (Salvage & Towage) Ltd, Piræus, Greece. Lagos registry closed. Registered at Piræus as **NISOS POROS**. 1968: Sold to Scrappingco NV, Antwerp, Belgium. 09.1968: Arrived Boom for breaking up. 09.1968: Registry closed.

NAME Official number Port Letters/Numbers	Yard No. Launched Registered	Registered		Engine Builder Horse Power Registered Speed	OWNER (Built for)
		G Ton N Ton	Length (feet) Beam (feet)		
FOSSA 180906 Steam tug	1316 05.03.1946 23.07.1946	66	70.5 17.1 7.2	British Polar 450bhp 6-cyl	Gaselee & Son Ltd, London

07.1946: Registered at London.   1960: Sold to Humphrey & Grey (Lighterage) Ltd, London.   Registered at London as **OWEN SMITH**.
1977: Re-engined with V8-cylinder 510bhp oil engine (D346) by Caterpillar Inc, Peoria, Illinois.   11.1982: Company wound up. Offered for sale.   1983: Sold to William F Mayhew, Queenborough. Registered at London as **FOSSA**.   1991: Sold for conversion to a houseboat. Registry cancelled.   2015: Believed still in existence at Chelsea.

Not built	1317, 1318 Orders cancelled				

MILFORD VISCOUNT 165644 M196 Motor trawler	1319 04.04.1946 10.01.1947	314 116	143.1 24.6 11.9	Ruston & Hornsby Milford Steam 750bhp 8-cyl 11.7 knots	Milford Steam Trawling Co Ltd, Milford Haven

Completed at a cost of £54,000; delivery delayed by engine makers and weather conditions.   04.07.1947: Registered at Milford (M196). 12.1949: On West of Scotland grounds in severe weather, trawl fouled propeller and fractured two blades. Steam trawler MILFORD KING (M226) responded to distress call but in weather, her bunkers shifted and a list developed. Both vessels managed to reach Oban safely. 29.03.1950: Sailed Milford for West of Ireland grounds (Sk Alex Smith); thirteen total crew.   30.03.1950: Sailed Castletownbere, Co Cork, for the Porcupine Bank grounds about 130 miles off the west coast of Cork.   01.04.1950: Seen by steam trawler DAMITO (LO81) (275grt/1917) fishing in position 52.30N 12.20W.   02.04.1950: At 7.30pm in R/T communication with DAMITO and MILFORD DUKE (M3) (see yard no.1346) and reported laid to due to heavy weather. No reliable record of any message after this time.   18.04.1950: Search of area by fishing vessels, RN, RAF & USAF. Nothing found.   20.04.1950: A lifebuoy was washed ashore near Ballybunion, Co Kerry. 25.04.1950: Search abandoned, though spurious messages purporting to be from the MILFORD VISCOUNT were received until 05.05.1950 and all investigated.   24.05.1950: Milford registry closed.   01.12.1950: At Board of Trade formal Inquiry at Milford, the Court found that the loss of the trawler and her thirteen crew was most probably due to an exceptional combination of weather and sea conditions. 04.04.1953: Wreckage trawled up in position 52.30N 14.40W and identified as from MILFORD VISCOUNT

INGÓLFUR ARNARSON RE201 Steam trawler	1320 13.05.1946 28.01.1947	654 243	177.2 30.0 15.0	Amos & Smith 1000ihp 3-cyl 12.0 knots	Bæjarútgerð Reykjavíkur, Reykjavik, Iceland

Oil fired.   Fitted with radar, possibly the first fishing vessel in the world to have this new electronic aid to navigation (Cossor Mk.1). 17.02.1947: Arrived in Reykjavík to an official welcome, as the first of 32 new trawlers ordered from British yards by the Icelandic government as the lynchpin of planned modernisation of Icelandic fisheries.   Later the vessels were suspected of structural weakness amidships in way of the trawl winch and following the loss of ELLIÐI (yard no.1325) in 1962, the hulls of all surviving sister ships were strengthened amidships. 02.1947: Registered at Reykjavik (RE201).   1972: Registered at Reykjavik as **HJÖRLEIFUR** (RE211).   02.12.1948: In very poor visibility aided by her radar, responded to distress call from Icelandic trawler JÚNÍ (GK345) (see yard no.736) stranded at Sauðanes between Súgandafjörður and Önundarfjörður due to faulty echo sounders. Rescued twenty-six crewmen by shooting a line from a lifeboat. Three remaining crewmen rescued from shore.   Sk Hannes Pálsson and eleven crewmen received a special commendation from the Icelandic Lifesaving Association for their part in the rescue.   18.01.1951: In a severe storm went to the assistance of the Norwegian motor ship TATRA (4766grt/1937) which was adrift with engine damage north of Orkney. Connected and delivered to Kirkwall.   03.12.1974: Sold to Desguaces y Recuperacion del Sur S.A., San Esteban de Pravia, Spain, for breaking up.   1974: Reykjavik registry closed. 11.02.1975: Breaking commenced.

HELGAFELL RE280 Steam trawler	1321 15.06.1946 11.03.1947	654 243	177.2 30.1 15.0	Amos & Smith 1000ihp 3-cyl 12.0 knots	Helgafell h/f, Reykjavik, Iceland

Oil fired.   08.1947: Registered at Reykjavik (RE280).   02.02.1953: Sold to Útgerðarfélag Akureyringa h/f, Akureyri.   1953: Reykjavik registry closed.   Registered at Akureyri as **SLÉTTBAKUR** (EA4).   18.04.1974: Sold for breaking up to Manuel Hevia Gonzales, Gijon, Spain, and arrived Gijon in tow of KALDBAKUR (see yard no.1322) also for breaking up.   1974: Akureyri registry closed.   03.10.1974: Breaking commenced.

KALDBAKUR EA1 Steam trawler	1322 30.07.1946 24.04.1947	654 242	177.2 30.2 15.0	Amos & Smith 1000ihp 3-cyl 12.0 knots	Útgerðarfélag Akureyringa h/f, Akureyri, Iceland

Oil fired.   05.1947: Registered at Akureyri (EA1).   18.04.1974: Sold for breaking up to Manuel Hevia Gonzales, Gijon, Spain and arrived Gijon towing SLÉTTBAKUR (see yard no.1321) also for breaking up.   1974: Akureyri registry closed.   27.05.1974: Breaking commenced.

EGILL SKALLAGRÍMSSON RE165 Steam trawler	1323 13.09.1946 03.06.1947	654 242	177.2 30.0 15.0	Amos & Smith 1000ihp 3-cyl 12.0 knots	h/f Kveldúlfur, Reykjavik, Iceland

Oil fired.   06.1947: Registered at Reykjavik (RE165).   21.01.1949: Rescued the seven crew members of the Icelandic fishing boat GUNNVÖR (RE 81) (102grt/1925) which was aground at Kögur, north-west coast of Iceland. Skipper Kolbeinn Sigurðsson and eleven crew members received a commendation from the Icelandic Lifesaving Association for the rescue.   22.09.1971: Sold to Jón Hafdal & Haraldur Jónsson, Hafnarfjörður.   Reykjavik registry closed.   Registered at Hafnarfjörður as **HAMRANES** (GK21).   02.01.1972: Sold to Haraldur H Júlíusson, Hreiðar Júlíusson and Bjarni Rafn Guðmundsson, trading as "Útgerðarfélagið Valur sf." Reykjavík. Hafnarfjörður registry closed. Registered at Reykjavík (RE 165).   18.06.1972: Foundered 50 miles off Snæfellsnes following an explosion in the forward fish room. All twenty-two crew rescued by the Reykjavík trawler NARFI (RE13) (980grt/1960).   1972: Reykjavik registry closed.

Milford Viscount (1319) was sadly lost along with her crew of thirteen in 1950.

(Cochrane archive)

The Icelandic **Ingólfur Arnarson** (1320).

(Cochrane archive)

Egill Skallagrímsson (1323) was another trawler built for owners in Iceland.

(Cochrane archive)

BJARNI RIDDARI	1324	657	177.2	Amos & Smith	Akurgerði h/f,
	26.10.1946	234	30.2	1000ihp 3-cyl	Hafnarfjörður,
GK1 Steam trawler	19.08.1947		14.9	12.0 knots	Iceland

Oil fired.   08.1947: Registered at Hafnarfjörður (GK1).   28.09.1964: Sold to N D Lagoutis & Sons, Piræus, Greece. Hafnarfjörður registry closed. Registered at Piræus as **NIKOLAOS**.   1966: Converted to motor and fitted with 6 cylinder 1150bhp oil engine by Atlas-Mak Masch, Kiel, and converted to a refrigerated cargo vessel.   08.08.1969: South of Villa Cisneros, Western Sahara, in approximate position 22° 12'N 016° 49'W whilst on passage from Arrecife, Canary Islands, towards Piræus, lost her propeller and went ashore.   Crew abandoned and believed vessel came afloat and foundered without trace.

ELLIÐI	1325	657	177.2	Amos & Smith	Bæjarútgerð Siglufjarðar,
	25.11.1946	230	29.9	1000ihp 3-cyl	Siglufjörður,
SI1 Steam trawler	01.10.1947		15.2	12.0 knots	Iceland

Oil fired.   10.1947: Registered at Siglufjörður (SI 1).   10.02.1962: In gale force winds 25 miles north-west of Snæfellsnes, at 5.00pm. sprung a leak amidships, started to take in water and to settle. At midnight twenty-six crewmen rescued by the Reykjavik trawler JUPITER (RE161) (804grt/1957) five minutes before the ship foundered but two crew members lost when carried away in an inflatable dinghy.   1962: Siglufjörður registry closed.

JÚLÍ	1326	657	177.2	Amos & Smith	Bæjarútgerð Hafnarfjarðar,
	24.01.1947	237	30.2	1000ihp 3-cyl	Hafnarfjörður,
GK21 Steam trawler	04.11.1947		15.2	12.0 knots	Iceland

Oil fired.   11.1947: Registered at Hafnarfjörður (GK21).   02.04.1948: Inbound for Hafnarfjörður, came across a lifeboat with the crew of seventeen of the British trawler LORD ROSS (H496) (265grt/1935), which had run aground at Álftanes in a snow storm, when enroute to Reykjavík. (Sk Benedikt Ögmundsson was rewarded by Lord Line Ltd with a gold watch).   02.1959: Reported missing presumed to have foundered and lost in heavy weather and severe icing conditions on 08.02.1959, 180 miles north-east of Gander, Newfoundland. All thirty crew lost.   1959: Hafnarfjörður registry closed.

ISÓLFUR	1327	655	177.2	Amos & Smith	Bjólfur h/f,
	06.05.1947	237	30.1	1000ihp 3-cyl	Seyðisfjörður,
NS14 Steam trawler	27.11.1947		15.2	12.0 knots	Iceland

Oil fired.   11.1947: Registered at Seyðisfjörður (NS14).   06.03.1957: Sold to Fiskiðjuver, Seyðisfjarðar. Registered at Seyðisfjörður as **BRIMNES** (NS14).   07.1968: Sold to Hughes Bolckow Ltd, Blyth, for breaking up.   07.1968: Seyðisfjörður registry closed.

RINOVIA	1328	557	170.0	Amos & Smith	Rinovia Steam
166667	19.06.1947	197	29.5	1040ihp 3-cyl	Fishing Co Ltd,
GY527 Steam trawler	19.01.1948		14.25	12.5 knots	Grimsby

Oil fired.   08.1945: Ordered.   First post-war trawler ordered by Grimsby owners and first to be oil fired.   First British-registered trawler to be fitted with radar as an aid to navigation (Kelvin Hughes Type 1).

16.01.1948: Registered at Grimsby (GY527). Insured for £109,000.   29.01.1948: Sailed on first trip.   18.02.1948: At Grimsby landed 3,981 boxes, £13,243 gross.   1950: First British registered trawler to have an autopilot fitted.   12.1960: Company and vessels sold to Ross Group, Grimsby.   13.12.1960: Registered at Grimsby as **ROSS STALKER** (GY527).   24.07.1964: Registered anew following alteration of particulars and conversion to motor; fitted with 7-cylinder 1300bhp oil engine by Ruston & Hornsby Ltd, Lincoln - 12.5 knots; now 564grt, 200net. Re-valued £177,252.   31.03.1966: Sold to Queen Steam Fishing Co Ltd, Grimsby.   07.11.1966: Registered at Grimsby as **ROSS RESOLUTION** (GY527).   30.09.1967: Sold to Ross Trawlers Ltd, Grimsby.   09.04.1968: Transferred to operate from Hull; Hellyer Bros Ltd managers and in their colours. Book value £90,314.   01.07.1969: Became part of British United Trawlers fleet (B.U.T.). B.U.T. livery adopted. 1969: Made six trips to the Icelandic grounds in 72 days, the shortest trip was 8 days.   16.06.1970: Sold to Hellyer Bros Ltd, Hull. 24.02.1977: Last landing at Hull and laid up for disposal.   04.04.1978: Sold to Richard J Brooks, Ipswich.   20.07.1978: Registered at Grimsby as **DEBUT**, classified as a dive support and salvage vessel.   03.09.1978: Sailed Plymouth for West Indies and worked for two years in Caribbean in various roles.   27.05.1980: Transited Panama Canal.   08.09.1980: Anchored off Baie de Taiohue, Nuku Hiva Island. Spent the next four and a half years in the Pacific Islands in a variety of roles.   26.02.1985: Sailed Funafatu, Tuvalu Group, for Australia, mostly under jury sail due to lack of bunker oil and crewed by only Capt Brooks and his wife Mariana.   25.06.1985: Anchored inside the Great Barrier Reef. HMAS TOWNSVILLE (FCPB205) transferred 3 tonnes of bunker oil and **DEBUT** sailed for Cairns, Queensland.   At Cairns, approached to do film work but with financial backing lacking, overstayed the twelve month rule for foreign registered ships and forced to anchor 25 miles east of Bloomfield. 30.06.1987: In cyclone, parted anchor cable and fetched up on Emily Reef 30 miles south-east of Cooktown, Queensland. Master/owner remained onboard for more than three years.   11.05.1989: Registry closed.

MURELLA	1329	550	171.0	C D Holmes	Trident Steam
181322	19.07.1947	200	29.2	925ihp 3-cyl	Fishing Co Ltd,
H481 Steam trawler	25.11.1947		14.7	11.5 knots	Hull

Oil fired.   Contract price £81,920.   22.11.1947: Registered at Hull (H481). Insured for £112,000.   27.11.1947: Sailed on first trip. 10.05.1951: Sold to Loch Steam Fishing Co Ltd, Hull.   21.05.1951: Registered at Hull as **LOCH MOIDART** (H481).   1953: Made three trips to Greenland fishery.   03.05.1954: Company bought by Hellyer Bros Ltd, Hudson Brothers Trawlers Ltd and Thomas Hamling & Co Ltd in equal shares.   09.09.1955: Answered distress call from steam trawler DANIEL QUARE (GY279) stranded in thick fog and calm seas 7 miles WNW of Langanes Point, Iceland. Stood by while Hull trawler CAMILLA (H193) connected and attempted to re-float. This was unsuccessful and vessel abandoned. All nineteen crew rescued by Icelandic lifesaving team. Later Icelandic gunboat THOR (693grt/1951) connected and attempted to refloat but vessel broke in two.   24.05.1966: Sold to Hellyer Bros Ltd, Hull.   09.1966: Transferred within the Associated Fisheries Group to Wyre Trawlers Ltd, Fleetwood.   14.09.1966: Sold to Wyre Trawlers Ltd, Fleetwood.   29.09.1966: Hull registry closed. 30.09.1966: Registered at Fleetwood (FD97).   25.04.1968: Sold to Jos de Smedt, Belgium, for breaking up.   10.05.1968: Arrived Antwerp for breaking up.   30.05.1968: Fleetwood registry closed.

Rinovia (1328) at anchor.

(Jonathan Grobler collection)

Murella (1329) was sold and renamed **Loch Moidart** in 1951. She was photographed as she entered her home port of Hull.

(Author's collection)

JUNELLA	1330	550	171.0	C D Holmes	J Marr & Son Ltd,
181327	02.09.1947	200	29.2	925ihp 3-cyl	Fleetwood
H497 Steam trawler	12.01.1948		14.7	11.5 knots	

Oil fired. 03.01.1948: Registered at Hull (H497). Insured for £112,000. 14.01.1948: Sailed on first trip. 11.04.1949: Sold to Alsey Steam Fishing Co Ltd, Grimsby, for £129,500. 10.04.1949: Hull registry closed. 29.04.1949: Registered at Grimsby (GY28). 04.11.1949: Registered at Grimsby as **KIRKNES** (GY28). 11.1960: Company and assets taken over by Ross Group Ltd, Grimsby. 25.11.1960: Registered at Grimsby as **ROSS HUNTER** (GY28). 20.12.1963: Sold to Hughes Bolckow Ltd, Blyth, for breaking up. 22.12.1963: Arrived Blyth. 01.03.1964: Grimsby registry closed.

DAUNTLESS STAR	1331	133	96.8	Crossley Bros	Star Drift Fishing Co Ltd
166709	16.09.1947	42	21.1	350bhp 6-cyl	Lowestoft
LT371 Motor drifter	02.02.1948		9.5	9.5 knots	

07.01.1948: Registered at Lowestoft (LT371). 09.11.1949: Prunier Trophy winner. 30.08.1951: Sold to Boston Deep Sea Fishing & Ice Co Ltd, Fleetwood. 1951: Sold to Booth Fisheries Canadian Co Ltd, Winnipeg, Manitoba, Canada. 1952: Re-engined with 6-cylinder 375bhp reconditioned oil engine by Crossley Bros, Manchester; 133grt. 28.02.1952: Registered at Halifax, NS ('H'43). 09.05.1951: Lowestoft registry closed. 1954: Registered at Halifax as **RED DIAMOND IV** ('H'43). 1960: Converted to refrigerated cargo ship. Halifax registry closed. Registered at Panama. 1985: Vessel removed from Lloyd's Register of Shipping - "Continued existence in doubt".

SUNLIT WATERS	1332	133	96.8	Crossley Bros	Frederick E Catchpole
166711	02.10.1947	42	21.1	350bhp 6-cyl	Lowestoft
LT377 Motor drifter	03.03.1948		9.5	9.5 knots	

07.02.1948: Registered at Lowestoft (LT377). 29.05.1951: Sold to Boston Deep Sea Fishing & Ice Co Ltd, Fleetwood. 1952: Re-engined with 6-cylinder 440bhp oil engine by Crossley Bros Ltd, Manchester. 08.05.1952: Registered at Lowestoft as **BOSTON SWIFT** (LT377). 03.09.1954: Sold to Mercury Fisheries Ltd, Halifax, Nova Scotia, Canada. 21.09.1954: Lowestoft registry closed. 09.1954: Registered at Halifax, NS ('H'11). 13.10.1954: Arrived Halifax, NS, with drifter ACORN (KY194) in tow disabled with depleted bunkers on delivery voyage, Lowestoft-Halifax, NS. By end 1956: Owned by Minister of Trade & Industry, Halifax, NS. 03.1957: Sold to George Murray & Craig Stores (Aberdeen) Ltd, Aberdeen. Halifax registry closed. 06.03.1957: Registered at Aberdeen (A143). 30.04.1957: Registered at Aberdeen as **SWIFTBURN** (A143). 24.04.1958: Sold to Star Drift Fishing Co Ltd, Lowestoft. 24.04.1958: Aberdeen registry closed. 26.04.1958: Registered at Lowestoft (LT367). 16.10.1958: Registered at Lowestoft as **DAUNTLESS STAR** (LT367). 21.10.1959: Prunier Trophy winner. 1965: Company and ships taken over by Boston Group, Hull. 10.1966: On charter to Triton Salvage & Towage Co Ltd for oil industry related work in North Sea. 12.02.1969: Sold to John R Francis & Alexander Reid, Gardenstown, and Frank West, Fraserburgh. Used as a herring carrier working with purse seiner BDELLIUM (FR185); not a success and laid up. 19.06.1971: Sold to J R Hashim, Frostenden, Suffolk. 04.12.1973: Sold to Warbler Fishing Co Ltd, Lowestoft. Employed as offshore standby safety vessel. 04.08.1975: Sold to Richard J Brooks, Ipswich. Lowestoft fishing registry closed. Moved to Ipswich to fit out as a diving support vessel. 1976: Sailed Ipswich for United Arab Emirates. Sold to local operators and based Dubai. Converted for use as a water carrier for shipping and offshore installations. 21.12.1999: Lowestoft registry closed. 2008: Vessel removed from Lloyd's Register of Shipping - "Continued existence in doubt".

GROOTVLEI	1333	492	163.5	Amos & Smith	National Trawling &
179895	31.10.1947	184	28.2	950ihp 3-cyl	Fishing Co Ltd,
CTA55 Steam trawler	08.04.1948		14.8	11.0 knots	Cape Town, South Africa.

04.1948: Registered at Cape Town (CTA55). 1971: Sold to S A Metals, Cape Town. 01.02.1971: Commenced breaking up at Cape Town. Cape Town registry closed.

BOSTON HURRICANE	1334	555	167.6	Amos & Smith	Boston Deep Sea Fishing &
181333	13.12.1947	205	29.0	925ihp 3-cyl	Ice Co Ltd,
H568 Steam trawler	01.06.1948		14.7	12.0 knots	Fleetwood

Oil fired. Contract price £110,000. 31.05.1948: Registered at Hull (H568). Insured for £114,000. 05.06.1948: Sailed on first trip. 13.04.1949: Sold to East Riding Trawlers Ltd, Hull. 25.05.1949: Registered at Hull as **STELLA POLARIS** (H568). 02.1951: Company in liquidation. 13.02.1951: Sold to Hudson Bros (Trawlers) Ltd, Hull, for £131,810. 09.03.1951: Registered at Hull as **CAPE CROZIER** (H568). 05.02.1960: Company and assets sold to Ross Group, Grimsby. 23.04.1965: Last landing at Hull. Laid up for disposal. 05.1965: Sold to Arie Rijsdijk-Boss en Zonen, Hendrik-Ido-Ambacht, Netherlands, for breaking up. 02.06.1965: Hull registry closed. 03.06.1965: Arrived Hendrik-Ido-Ambacht.

ST. CHRISTOPHER	1335	555	167.5	Amos & Smith	Saint Andrew's Steam
181337	28.01.1948	205	29.0	925ihp 3-cyl	Fishing Co Ltd,
H573 Steam trawler	14.07.1948		15.25	12.0 knots	Hull

Oil fired. Contract price £110,000. 05.07.1945: Registered at Hull (H573). Insured for £114,000. 18.07.1948: Sailed on first trip. 14.04.1949: Sold to F & T Ross Ltd, Hull. 21.06.1949: Registered at Hull as **TESLA** (H573). 27.04.1950: Sailed for Greenland on a salting voyage. 12.07.1950: Landed at Lowestoft after 79 days, 194tons 2cwts, £8,250 gross. 19.04.1955: Sold to Derwent Trawlers Ltd, Grimsby for £86,507. 10.05.1955: Registered at Hull as **STELLA CARINA** (H573). 11.03.1959: Shortly after leaving Hull fish dock, in collision with anchored collier MENDIP (1362grt/1950). A gash 22 feet wide and 3 feet deep was ripped down the port side. Began to settle by the stern, Ch Eng George Roberts working waist deep in water remained at the controls; Sk Fred Sullivan turned the ship and drove onto the river bank close to Victoria Dock. Twenty crewmen picked up from the ship's life rafts by a pilot cutter. 27.05.1959: Re-floated and towed to Immingham for repairs. 25.03.1963: Sold to Hudson Brothers Trawlers Ltd, Hull. 01.10.1964: Sold to Alsey Steam Fishing Co Ltd, Hull. 01.01.1965: Owners restyled Hudson Fishing Co Ltd. 22.12.1965: Registered at Hull as **ROSS CARINA** (H573). Ross Group funnel colours adopted. 31.03.1966: Sold to Hudson Brothers Trawlers Ltd Hull. 22.11.1966: Landed and laid up for disposal. 02.1967: Sold to Scrappingco S.r.l., Brussels, Belgium. 15.02.1967: Hull registry closed. 11.03.1967: Arrived Willebroek for breaking up.

LORD WAVELL	1336	636	175.0	C D Holmes	Lord Line Ltd,
181346	25.03.1948	245	30.7	1000ihp 3-cyl	Hull
H578 Steam trawler	09.09.1948		15.1	12.0 knots	

26.04.1947: Contract signed for six trawlers to this design, yard Nos 1336 - 1341, at a cost of £99,000 each. Revised cost to build £104,457. Final cost fitted out £107,998.

07.09.1948: Registered at Hull (H583). 11.09.1948: Sailed on first trip. 14.12.1949: Completed inclining for new stability data after conversion for burning oil fuel by Humber St Andrew's Engineering Co Ltd, Hull. Total cost £13,173. 17.04.1950: Lord Line Ltd in liquidation. Sold to Associated Fisheries Trawling Co Ltd, Hull, for £79,000. 22.07.1953: Company restyled Lord Line Ltd, Hull.

08.04.1963: Last landing at Hull. 10.04.1963: Following reorganisation of the Associated Fisheries Group, Hull fleet transferred to Grimsby. Associated Fisheries funnel colours adopted. 22.04.1963: Hull registry closed. 23.04.1963: Registered at Grimsby (GY97). 22.02.1966: Sold to Northern Trawlers Ltd, Grimsby. 04.1967: Transferred within Associated Fisheries Group to Wyre Trawlers Ltd, Fleetwood. 26.04.1967: Sold to Wyre Trawlers Ltd, Fleetwood. 19.05.1967: Grimsby registry closed. 22.05.1967: Registered at Fleetwood (FD98). 05.1967: Sold to L E Adams, Plymouth, c/o Ocean Marine Salvage Co, Plymouth, for conversion to a diving and salvage vessel. 05.02.1968: Fleetwood fishing vessel registry closed. Unofficially renamed **LORD J. WAVELL** to satisfy terms of sale. 1968: Arrested by Admiralty Marshall at Falmouth. 1970: Vessel sold at auction to Haulbowline Industries Ltd, Passage West, Co Cork, for breaking up. 29.03.1970: Delivered Passage West from Falmouth in tow. 18.03.1975: Registry closed except in respect of mortgage - "Vessel broken up". 01.11.1978: Registry closed.

LORD ANCASTER	1337	636	175.0	Amos & Smith	Lord Line Ltd,
181350	24.04.1948	245	30.9	1050ihp 3-cyl	Hull
H583 Steam trawler	07.10.1948		15.1	12.0 knots	

Revised cost to build £104,200. Final cost fitted out £107,565.
28.09.1948: Registered at Hull (H583). 11.10.1948: Sailed on first trip. 17.04.1950: Lord Line Ltd in liquidation. Sold to Associated Fisheries Trawling Co Ltd, Hull, for £75,000. 03.05.1950: Completed inclining for new stability data after conversion for burning oil fuel by Humber St Andrew's Engineering Co Ltd, Hull. Total cost £12,459. 13.05.1952: Suffered extensive fire damage to bridge deck whilst in dry dock at Hull for special survey. 22.07.1953: Company restyled Lord Line Ltd, Hull. 1953: Made two trips to Greenland fishery.
01.10.1965: Transferred within Associated Fisheries Group to Hellyer Bros Ltd, Hull. 24.05.1966: Sold to Hellyer Bros Ltd, Hull for £91,821. 10.10.1967: Landed and laid up for disposal. 1967: Sold to Clayton & Davie Ltd, Dunston, for breaking up. 30.11.1967: Arrived River Tyne. 29.11.1968: Hull registry closed - "Vessel broken up".

LORD ROWALLAN	1338	636	175.0	C D Holmes	Lord Line Ltd,
183381	23.07.1948	244	30.7	1050ihp 3-cyl	Hull
H9 Steam trawler	26.01.1949		15.1	11.5 knots	

Oil fired. Revised cost to build £106,366. Final cost fitted out £109,106. 26.01.1949: Registered at Hull (H9). 29.01.1949: Sailed on first trip. 17.04.1950: Lord Line Ltd in liquidation. Sold to Associated Fisheries Trawling Co Ltd, Hull, for £75,000. 22.07.1953: Company restyled Lord Line Ltd, Hull. 1953: Made two trips to Greenland fishery. 16.04.1963: Last landing at Hull. 10.04.1963: Following reorganisation of Associated Fisheries Group, Hull fleet transferred to Grimsby. Associated Fisheries funnel colours adopted. 22.04.1963: Hull registry closed. 23.04.1963: Registered at Grimsby (GY98). 22.02.1966: Sold to Northern Trawlers Ltd, Grimsby. 11.1968: Sold to Van Heyghen Frères, Ghent, Belgium, for breaking up. 22.11.1967: Grimsby registry closed. 30.11.1967: Arrived Ghent.

LORD WILLOUGHBY	1339	635	177.9	Amos & Smith	Lord Line Ltd,
183388	20.09.1948	243	30.7	1050ihp 3-cyl	Hull
H36 Steam trawler	22.02.1949		15.1	12.2 knots	

Oil fired. Revised cost to build £106,421. Final cost fitted out £109,245. 23.02.1949: Registered at Hull (H36). 25.02.1949: Sailed on first trip. 17.04.1950: Lord Line Ltd in liquidation. Sold to Associated Fisheries Trawling Co Ltd, Hull, for £75,000. 22.07.1953: Company restyled Lord Line Ltd, Hull. 1953: Made one trip to Greenland fishery. 17.04.1963: Last landing at Hull. 10.04.1963: Following reorganisation of Associated Fisheries Group, Hull fleet transferred to Grimsby. Associated Fisheries funnel colours adopted. 22.04.1963: Hull registry closed. 23.04.1963: Registered at Grimsby (GY102). 22.02.1966: Sold to Northern Trawlers Ltd, Grimsby. 10.1968: Sold to Jos de Smedt, Antwerp, for breaking up. 08.10.1968: Arrived Antwerp. 21.10.1968: Grimsby registry closed.

LORD FRASER	1340	635	177.9	C D Holmes	Lord Line Ltd,
183392	03.11.1948	243	30.7	1050ihp 3-cyl	Hull
H48 Steam trawler	12.04.1949		15.1	12.2 knots	

Oil fired. Revised cost to build £108,478. Final cost fitted out £111,234. 12.04.1949: Registered at Hull (H48). 14.04.1949: Sailed on first trip. 17.04.1950: Lord Line Ltd in liquidation. Sold to Associated Fisheries Trawling Co Ltd, Hull, for £75,000. 22.07.1953: Company restyled Lord Line Ltd, Hull. 02.04.1963: Last landing at Hull. 05.04.1963: Following reorganisation of Associated Fisheries Group, Hull fleet transferred to Grimsby. Associated Fisheries funnel colours adopted. 22.04.1963: Hull registry closed. 23.04.1963: Registered at Grimsby (GY108). 22.02.1966: Sold to Northern Trawlers Ltd, Grimsby. 04.1968: Sold to Jos de Smedt, Antwerp, Belgium, for breaking up. 04.04.1968: Grimsby registry closed. 19.04.1968: Arrived Antwerp.

LORD CUNNINGHAM	1341	635	177.9	Amos & Smith	Lord Line Ltd,
183396	18.12.1948	243	30.7	1050ihp 3-cyl	Hull
H69 Steam trawler	17.05.1949		15.1	12.2 knots	

Oil fired. Revised cost to build £110,533. Final cost fitted out £111,248. 15.05.1949: Registered at Hull (H36). 19.05.1949: Sailed on first trip. 17.04.1950: Lord Line Ltd in liquidation. Sold to Associated Fisheries Trawling Co Ltd, Hull, for £75,000. 22.07.1953: Company restyled Lord Line Ltd, Hull. 10.04.1963: Last landing at Hull. 16.04.1963: Following reorganisation of Associated Fisheries Group, Hull fleet transferred to Grimsby. Associated Fisheries funnel colours adopted. 22.04.1963: Hull registry closed. 23.04.1963: Registered at Grimsby (GY109). 20.06.1963: Stranded on the reef at the east end of the island of Fugløya, Norway, close to Rotvaer Lighthouse. Refloated by local tugs and taken to Lødingen for repair before returning home. 22.02.1966: Sold to Northern Trawlers Ltd, Grimsby. 11.1967: Sold to Van Heyghen Frères, Ghent, Belgium, for breaking up. 22.11.1967: Grimsby registry closed. 30.11.1967: Arrived Ghent.

Boston Hurricane (1334)

(Author's collection)

Lord Wavell (1336)

(Author's collection)

Lord Rowallan (1338)

(Author's collection)

ERNEST HOLT	1342	599	177.6	Amos & Smith	Ministry of Agriculture
182627	09.06.1948	182	30.2	900ihp 3-cyl	and Fisheries,
GY591 Fishery research	16.12.1948		15.1	11.5 knots	London
ship					

Oil fired.   11.12.1948: Registered at Grimsby (GY521).   1955: Transferred to the Ministry of Agriculture, Fisheries & Food, London.
19.01.1970: Transferred to the Department of Agriculture & Fisheries for Scotland, Leith, Edinburgh.   04.03.1971: Grimsby registry closed.
Converted for fishery patrol duties.   06.71: Registered at Leith as **SWITHA**.   31.01.1980: Inwards for Leith in severe weather conditions
stranded on Herwit Rocks SE of Inchkeith, Firth of Forth. All twenty-five crew taken off by helicopter.   07.02.1980: Salvage proved
impossible and wreck was opened up by explosive charges placed by Royal Navy divers to allow fuel oil to be recovered and wreck
subsequently dispersed.   Registry closed.

BOYNTON WYKE	1343	676	178.4	Amos & Smith	West Dock Steam Fishing
183403	14.02.1948	268	30.6	1050ihp 3-cyl	Co Ltd,
H74 Steam trawler	30.06.1949		15.2	12.2 knots	Hull

05.09.1947: Contract signed cost to build £98,432.   Final cost fitted out £106,322. The last coal burning steam trawler built for Hull owners.
30.06.1949: Registered at Hull (H74).   03.07.1949: Sailed on first trip.   24.11.1952: Homeward from the Greenland grounds developed
boiler trouble and towed by another Hull trawler to Aberdeen for repairs.   1953: Made two trips to Greenland fishery.   07.07.1954: Company
and assets sold to Associated Fisheries Trawling Co Ltd, Hull, and Boyd Line Ltd, Hull, joint shareholders.   15.07.1955: Converted for
burning oil fuel at total cost £17,500.   14.05.1959: Sold to Boyd Line Ltd, Hull, for £126,410.   22.05.1959: Registered at Hull as **ARCTIC
CRUSADER** (H74).   22.10.1969: Last landing at Hull. Laid up for disposal.   11.1969: Sold to Hughes Bolckow Ltd, Blyth, for breaking up.
10.11.1969: Arrived Blyth.   12.11.1969: Hull registry closed - "Vessel broken up".

| **Not built** | 1344, 1345   Orders cancelled | | | | |

MILFORD DUKE	1346	363	145.1	Ruston (Vickers, Barrow)	Milford Steam Trawling
183931	15.03.1949	127	25.7	1050ihp 8-cyl	Co Ltd,
M3 Motor trawler	12.09.1949		12.8	11.5 knots	Milford Haven

13.09.1949: Registered at Milford (M3).   03.1955: Sold to Soc Anon d'Armement Mallett, Dieppe, France.   31.03.1955: Milford registry
closed. Registered at Dieppe as **JEAN-VAUQUELIN** (DI.1593).   1968: Sold to Claridge Trawlers Ltd, Lowestoft.   13.07.1968: Arrived
Lowestoft towing **JOLI FRUCTIDOR** (DI.1585) (see yard no.1347). Dieppe registry closed.   25.09.1968: Registered at Lowestoft as
ST. ROSE (LT82).   30.09.1968: First landing at Lowestoft.   1973: Re-engined with 6-cylinder 1350bhp oil engine by Ruston Paxman
Diesels Ltd, Newton-le-Willows.   1984: Last year fishing grossed £247,178.   1985: Decommissioning grant paid.   03.1985: Stripped of all
usable gear.   28.03.1985: Registered at Lowestoft as **UNDA** to free name. Registered as offshore platform standby/safety vessel.
24.05.1985: Main engine removed.   06.1985: Sold to Swanscombe Ltd, Erith. for breaking up.   03.07.1985: Sailed Lowestoft for Gravesend
in tow of motor tug MICHEL PETERSEN (195grt/1964) along with former trawler BARBADOS (LT312) (213grt/1958) also for breaking.
02.01.1986: Lowestoft registry closed.

MILFORD DUCHESS	1347	363	145.1	Ruston (Vickers, Barrow)	Milford Steam Trawling
183932	30.03.1949	127	25.7	1050ihp 8-cyl	Co Ltd,
M16 Motor trawler	11.10.1949		12.8	11.5 knots	Milford Haven

03.10.1949: Registered at Milford (M16).   03.1955: Sold to Soc d'Armement Mallett, Dieppe, France.   28.06.1955: Milford registry closed.
Registered at Dieppe as **JOLI FRUCTIDOR** (DI.1585).   1968: Sold to Claridge Trawlers Ltd, Lowestoft.   13.07.1968: Arrived Lowestoft in
tow of **JEAN-VAUQUELIN** (DI.1593) (see yard no.1346) with main engine on deck.   1968: Dieppe registry closed.   12.1968: Registered at
Lowestoft as **ST. NICOLA**.   1970: Re-engined with 6-cylinder 1350bhp oil engine by Ruston & Hornsby Ltd, Lincoln.
26.12.1970: Registered at Lowestoft as **ST. NICOLA** (LT83).   02.01.1971: Landed first trip at Lowestoft.   Late 1984: Laid up at Lowestoft.
1985: Decommissioning grant paid.   02.1985: Main engine removed.   03.1985: Registered at Lowestoft as **WILLEM ADRIANA** to free
name.   Registered as offshore platform standby/safety vessel.   04.1985: Sold to Tidal Marine & Dredging Services (Essex) Ltd, Grays,
Essex, for breaking up.   24.04.1985: Sailed Lowestoft for Gravesend in tow of motor tug MICHEL PETERSEN (195grt/1964) along with
former trawler BURNBANKS (217grt/1959) also for breaking.   02.01.1986: Lowestoft registry closed.

NETHE	1348	100	75.0	Plenty & Son	Soc Générale de Dragage,
	14.05.1949		19.1	900ihp 3-cyl	Antwerp,
Steam tug	22.09.1949		9.3	9.5 knots	Belgium

09.1949: Registered at Antwerp.   1968: Removed from Lloyd's Register of Shipping.

SENNE	1349	100	75.0	Plenty & Son	Soc Générale de Dragage,
	14.05.1949		19.1	900ihp 3-cyl	Antwerp,
Steam tug	13.10.1949		9.3	9.5 knots	Belgium

10.1949: Registered at Antwerp.   1965: Re-engined with 6-cylinder 500bhp oil engine by Motorenwerk Mannheim A G, Mannheim - 10knots.
1992: Removed from Lloyd's Register of Shipping.

FLATHOLM	1350	351	140.8	Amos & Smith	Neale & West Ltd,
169546	11.06.1949	134	25.1	605ihp 3-cyl	Cardiff
CF23 Steam trawler	16.11.1949		11.9	11.0 knots	

Oil fired.   11.1949: Registered at Cardiff (CF23).   10.1950: Sold to National Sea Products Ltd, Halifax, Nova Scotia.   11.1950: Sailed Hull
(Capt R A Winn) with future Canadian master Frank Tidman also onboard.   19.11.1950: On delivery voyage stranded in dense fog
approximately 28 miles south of St John's, Newfoundland. Became a total loss.   12.1950: Cardiff registry closed. (Was to have been
renamed CAPE SAMBRO under Canadian registration).

Lord Fraser (1340)

(Author's collection)

Boynton Wyke (1343)

(Cochrane archive)

The tug **Ala** assists the **St Rose** (1346) into Lowestoft. This trawler was built as **Milford Duke**.

(Author's collection)

GATEHOLM	1351	351	140.8	Amos & Smith	Neale & West Ltd,
169547	10.08.1949	134	25.1	605ihp 3-cyl	Cardiff
CF24 Steam trawler	15.12.1949		11.9	11.0 knots	

Oil fired. 12.1950: Registered at Cardiff (CF24). 1950: Sold to National Sea Products Ltd, Halifax, Nova Scotia. 31.10.1950: Arrived Halifax via St John's, Newfoundland (Capt Frank Green). Cardiff registry closed. Registered at Halifax as **CAPE SMOKY** (Canadian Fisheries No. 55). 1968: Sold to Ocean Fisheries Ltd, Halifax. 1968: Sold to shipbreakers and broken up. Halifax registry closed.

ANDANES	1352	724	180.0	Amos & Smith	Rinovia Steam Fishing
182643	26.08.1949	254	31.0	1290ihp 3-cyl	Co Ltd,
GY53 Steam trawler	08.02.1950		16.0	16.0 knots	Grimsby

Oil fired. Fitted with autopilot and radar (Sperry Single-Unit Gyro Pilot and Kelvin Hughes Type 1A series II radar). 02.1950: Registered at Grimsby (GY53). Insured for £147,500. 05.1950: Sold to Andanes Steam Fishing Co Ltd, Grimsby. 01.07.1960: Company and assets sold to Ross Group, Grimsby. 08.1961: Registered at Grimsby as **ROSS FIGHTER** (GY53). 07.10.1963: New certificate issued following alteration of particulars. Re-engined with 8-cylinder 1306bhp oil engine by Ruston & Hornsby Ltd, Lincoln - 12.5 knots; 761grt 294net. Fitted with a battery of Jackstone vertical plate freezers on main deck thus becoming Britain's first freezer trawler. Most of work completed by Ross Group Engineers, Grimsby, at a cost of £37,00 with WFA grant. Re-valued at £239,000. 05.1966: Registered at Grimsby as **ROSS RAMILLES** (GY53). 10.1966: Sold to Ross Trawlers Ltd, Grimsby. Book value £92,268. 21.05.1979: Sold to David Potts (Steel) Ltd, Stallingborough, for breaking up at the former Doig yard at Grimsby. 13.12.1979: Grimsby registry closed.

NEWBY WYKE	1353	672	178.4	Amos & Smith	West Dock Steam Fishing
183429	08.10.1949	253	30.6	1000ihp 3-cyl	Co Ltd,
H111 Steam trawler	09.03.1950		14.8	11.7 knots	Hull

Oil fired. 05.09.1947: Contract signed - cost to build £102,808. Final cost fitted out £113,637. The last vessel ordered by the company. 10.03.1950: Registered at Hull (H111). 11.03.1950: Sailed on first trip. 1953: Made three trips to Greenland fishery. 07.07.1954: Company and assets sold to Associated Fisheries Trawling Co Ltd, Hull, and Boyd Line Ltd, Hull, joint shareholders. 15.04.1959: Transferred to Lord Line Ltd whose funnel colours adopted. 01.10.1965: Transferred within the Associated Fisheries Group to Hellyer Bros Ltd, Hull, for £71,266. Hellyer livery adopted. 24.05.1966: Sold to Hellyer Bros Ltd, Hull. 01.07.1969: Became part of the British United Trawler (B.U.T.) fleet. B.U.T. livery adopted. Book value £17,025. 30.01.1975: Last landing. Laid up for disposal. 23.04.1975: Sold to Albert Draper & Son Ltd, Hull, for breaking up. 15.07.1975: Breaking commenced at the Victoria Dock slipway. 07.08.1975: Hull registry closed - "Vessel broken up"

NORTHERN CHIEF	1354	692	181.6	Amos & Smith	Northern Trawlers Ltd,
182647	22.11.1949	246	31.2	1100ihp 3-cyl	Grimsby
GY128 Steam trawler	04.05.1950		16.0	12.5 knots	

Oil fired. Cost to build £130,701. 05.05.1950: Registered at Grimsby (GY128). 01.07.1969: Became part of the British United Trawler (B.U.T.) fleet. B.U.T. livery adopted. Book value £17,500. 27.01.1975: Last landing. Laid up for disposal. 21.04.1975: Company restyled British United Trawlers Ltd. 24.06.1975: Sold to Blyth Shipbreakers & Repairers Ltd, Blyth, for breaking up in an en bloc deal with NORTHERN SEA (see yard no.1355) and NORTHERN ISLES (see yard no.1356). 19.05.1976: Grimsby registry closed - "Vessel broken up".

NORTHERN SEA	1355	692	181.6	Amos & Smith	Northern Trawlers Ltd,
182652	19.01.1950	246	31.2	1100ihp 3-cyl	Grimsby
GY142 Steam trawler	20.06.1950		16.0	12.5 knots	

Oil fired. Cost to build £130,428. 21.06.1950: Registered at Grimsby (GY142). 01.07.1969: Became part of the British United Trawler (B.U.T.) fleet. B.U.T. livery adopted. Book value £17,500. 27.01.1975: Last landing. Laid up for disposal. 21.04.1975: Company restyled British United Trawlers Ltd. 24.06.1975: Sold to Blyth Shipbreakers & Repairers Ltd, Blyth, for breaking up in an en bloc deal with NORTHERN CHIEF (see yard no.1354) and NORTHERN ISLES (see yard no.1356). 19.05.1976: Grimsby registry closed - "Vessel broken up".

NORTHERN ISLES	1356	692	181.6	Amos & Smith	Northern Trawlers Ltd,
182658	06.03.1950	246	31.2	1100ihp 3-cyl	Grimsby
GY149 Steam trawler	17.08.1950		16.0	12.5 knots	

Oil fired. Cost to build £133,688. 17.08.1950: Registered at Grimsby (GY149). 01.07.1969: Became part of the British United Trawler (B.U.T.) fleet. B.U.T. livery adopted. Book value £17,500. 23.01.1975: Last landing. Laid up for disposal. 21.04.1975: Company restyled as British United Trawlers Ltd. 24.06.1975: Sold to Blyth Shipbreakers & Repairers Ltd, Blyth, for breaking up in an en bloc deal with NORTHERN CHIEF (see yard no.1354) and NORTHERN SEA (see yard no.1355). 18.05.1976: Grimsby registry closed - "Vessel broken up".

MILFORD KNIGHT	1357	168	107.8	Ruston V A Barrow	Milford Steam Trawling
183934	04.04.1950	63	21.1	420bhp 8-cyl	Co Ltd,
M127 Motor trawler	14.09.1950		10.3	11.2 knots	Milford Haven

12.09.1950: Registered at Milford (M127). 22.12.1954: Whilst sheltering in Tobermory Bay (Sk Jack Clarke), driven ashore. 23.12.1954: Swansea trawler HARLECH CASTLE (SA42) connected and successfully refloated. 09.1955: Sold to Clan Steam Fishing Co (Grimsby) Ltd, London Colney, Hertfordshire. 10.10.1955: Milford registry closed. 13.10.1955: Registered at Lowestoft as **TRINIDAD** (LT210). 25.03.1957: Sold to Drifter Trawlers Ltd, Lowestoft. 1976: Engine removed. 1976: Sold to T G Darling, Oulton Broad, Lowestoft, for breaking up. 05.1976: Delivered Oulton Broad for breaking up by East Anglian Reclamation Ltd at Nelson Wharf. 1976: Lowestoft registry closed - "Vessel broken up".

MILFORD COUNTESS	1358	168	107.8	Ruston V A Barrow	Milford Steam Trawling
183933	24.05.1950	63	21.1	420bhp 8-cyl	Co Ltd,
M128 Motor trawler	14.11.1950		10.3	11.5 knots	Milford Haven

16.11.1950: Registered at Milford (M128).   03.1955: Sold to Colne Fishing Co Ltd, Lowestoft.   22.03.1955: Milford registry closed.
26.04.1955: Registered at Lowestoft as **TOBAGO** (LT182).   19.03.1964: Returning to Lowestoft (Sk E E Whitcombe, Grimsby), in heavy weather driven aground at entrance, abandoned by crew and turned on her port side.   29.04.1964: Refloated after sand and shingle removed. Beyond economical repair, engine removed and laid up pending sale.   09.1964: Sold to Lacmots Ltd for breaking up.
08.10.1964: Arrived Queenborough in tow of TOGO (LT69) (76grt/1905) also for breaking up.   17.01.1967: Lowestoft fishing registry closed.
16.02.1967: Lowestoft registry closed.

BRADMAN	1359	693	180.0	Amos & Smith	Bunch Steam Fishing
182662	01.07.1950	252	31.0	1200ihp 3-cyl	Co Ltd,
GY161 Steam trawler	11.12.1950		16.0	16.0 knots	Grimsby

Oil fired.   Cost to build £144,222.   24.11.1950: Registered at Grimsby (GY161).   01.03.1965: Company and assets sold to Ross Group, Grimsby.   26.11.1965: Registered at Grimsby as **ROSS ANSON** (GY161).   01.01.1967: Sold to Hudson Bros (Trawlers) Ltd, Hull (Ross Group).   02.1967: At Harstad, Norway, (Sk Harry Saunderson) connected to disabled Hull steam trawler ROSS PROCYON (H184) (590grt/1950) (Sk Raymond A Jopling) and delivered her off the Humber Lightship after a tow of 1,130 miles.   23.01.1967: Both ships landed at Hull.   28.06.1967: Landed at Hull. Laid up for disposal.   07.1968: Sold to Jos de Smedt, Antwerp, Belgium.   20.07.1968: Sailed Hull for Antwerp in tow of motor tug FAIRPLAY II (136grt/1959) in tandem tow wth ST ARCADIUS (H207) (576grt/1951).
01.08.1968: Grimsby registry closed.

LORD LOVAT	1360	713	181.3	Amos & Smith	Associated Fisheries
183464	15.08.1950	256	31.1	1000ihp 3-cyl	Trawling Co Ltd,
H148 Steam trawler	30.01.1951		15.1	13.0 knots	Hull

Oil fired. 31.01.1948: Contract signed for six trawlers to this design, yard Nos.1360-1365 for Lord Line Ltd, Hull.   17.04.1950: Lord Line in liquidation having made payments of £34,932. Contract taken over by Associated Fisheries Trawling Co Ltd, Hull.   Cost to build: £127,066.   Final cost fitted out, £129,938.
31.01.1951: Registered at Hull (H148).   03.02.1951: Sailed on first trip.   22.07.1953: Company restyled Lord Line Ltd, Hull.   1953: Made one trip to Greenland fishery.   01.10.1965: Transferred within Associated Fisheries Group to Hellyer Bros Ltd, Hull. Associated - Hellyer livery adopted.   24.05.1966: Sold to Hellyer Bros Ltd, Hull, for £132,285.   01.07.1969: Ross Group amalgamated with Associated Fisheries to form British United Trawlers Ltd, Hull. B.U.T. funnel colours adopted. Book value £24,425.   13.08.1975: Landed. Laid up for disposal.
08.03.1976: Sold to Shipbreaking (Queenborough) Ltd, Queenborough, and allocated to Cairnryan for breaking up.   12.04.1976: Sailed Hull for Cairnryan.   14.07.1976: Completed breaking up.   08.11.1976: Hull registry closed - "Vessel broken up".

LORD TEDDER	1361	722	181.3	C D Holmes	Associated Fisheries
183469	14.09.1950	260	31.0	1050ihp 3-cyl	Trawling Co Ltd,
H154 Steam trawler	06.03.1951		15.1	12.2 knots	Hull

Oil fired.   31.01.1948: Contract signed with Lord Line Ltd, Hull.   17.04.1950: Lord Line in liquidation having made payments of £12,685.
Contract taken over by Associated Fisheries Trawling Co Ltd, Hull.   Cost to build £129,639.   Final cost fitted out £130,420.
07.03.1951: Registered at Hull (H154).   09.03.1951: Sailed on first trip.   22.07.1953: Company restyled Lord Line Ltd, Hull.   1953: Made one trip to Greenland fishery.   01.10.1965: Transferred within Associated Fisheries Group to Hellyer Bros Ltd, Hull. Associated - Hellyer livery adopted.   24.05.1966: Sold to Hellyer Bros Ltd, Hull, for £132,285.   01.07.1969: Ross Group amalgamated with Associated Fisheries to form British United Trawlers Ltd, Hull. B.U.T. funnel colours adopted. Book value £24,462.   17.07.1975: Landed. Laid up for disposal.
08.03.1976: Sold to Shipbreaking (Queenborough) Ltd, Queenborough, and allocated to Cairnryan for breaking up.
13.08.1976: Commenced breaking up at Cairnryan.   08.11.1976: Hull registry closed - "Vessel broken up".

LORD MOUNTEVANS	1362	712	181.3	Amos & Smith	Associated Fisheries
183473	11.11.1950	256	31.1	1000ihp 3-cyl	Trawling Co Ltd,
H169 Steam trawler	26.04.1951		15.1	13.0 knots	Hull

Oil fired.   31.01.1948: Contract signed with Lord Line Ltd, Hull.   17.04.1950: Lord Line in liquidation having paid a deposit of £5,000.
Contract taken over by Associated Fisheries Trawling Co Ltd, Hull.   Cost to build £129,596.   Final cost fitted out £133,525.
27.04.1951: Registered at Hull (H169).   30.04.1951: Sailed on first trip.   22.07.1953: Company restyled Lord Line Ltd, Hull.   1953: Made two trips to Greenland fishery.   20.03.1963: Last landing at Hull.   11.04.1963: Following reorganisation of Associated Fisheries Group, Hull fleet transferred to Grimsby. Associated Fisheries funnel colours adopted.   22.04.1963: Hull registry closed.   23.04.1963: Registered at Grimsby (GY79).   01.10.1965: Transferred within Associated Fisheries Group to Hellyer Bros Ltd for £139,048.   22.02.1966: Sold to Northern Trawlers Ltd, Grimsby, for £139,048.   01.07.1969: Ross Group amalgamated with Associated Fisheries to form British United Trawlers Ltd, Hull. B.U.T. funnel colours adopted; book value £25,199.   03.10.1973: Sold to Albert Draper & Son Ltd, Hull, for breaking up.
15.11.1973: Breaking commenced at Victoria Dock slipway.   12.03.1974: Grimsby registry closed.

Not built	1363, 1364, 1365. 24.01.1950: Contract cancelled. Shipyard refunded £5,000 deposit on all 3 vessels.

NORTHERN JEWEL	1366	799	185.5	Amos & Smith	Northern Trawlers Ltd,
184920	26.09.1953	313	32.2	1100ihp 3-cyl	London
GY1 Steam trawler	01.06.1954		17.0	12.5 knots	

Oil fired.   Keel laid for Rinovia Steam Fishing Co Ltd, Grimsby, as KOPANES; sold on the stocks to Northern Trawlers Ltd, London. Cost to build £197,789.
02.06.1954: Registered at Grimsby (GY1).   01.07.1969: Became part of the British United Trawler (B.U.T.) fleet. B.U.T. livery adopted.
Book value £60,125.   29.01.1975: Last landing. Laid up for disposal.   21.04.1975: Company restyled British United Trawlers Ltd.
03.03.1976: Sold to H Kitson Vickers (Engineers) & Sons Ltd, Sheffield, for breaking up in an en bloc deal with NORTHERN SCEPTRE (see yard no.1387).   14.05.1976: Resold to London Demolition Co Ltd and allocated to Medway Secondary Metals Ltd, Rainham, Kent, and broken up.   19.03.1979: Grimsby registry closed.

Northern Sea (1355)

(David Sllinger collection)

Milford Knight (1357)

(Cochrane archive)

Lord Lovat (1360) outward bound from Hull.

(Author's collection)

ARCTIC WARRIOR	1367	712	181.3	Amos & Smith	Boyd Line Ltd,
185086	25.01.1951	256	31.1	1300ihp 3-cyl	Hull
H176 Steam trawler	18.06.1951		15.1	13.0 knots	

Oil fired. 31.01.1948: Contract signed.
19.06.1951: Registered at Hull (H176). Insured for £161,000. 21.06.1951: Sailed on first trip. 1953: Made four trips to Greenland fishery including one trip to Labrador. 1954: First winner of the Silver Cod Trophy (awarded by the British Trawler Federation at London to the skipper and crew with the largest total catch for the year) with skippers John William Hamling and Richard Sackville-Bryant - 42,776 x 10 stone kit; 332 days at sea; £123,525 gross. 1956: Silver Cod Trophy runner-up, skipper Richard Sackville-Bryant - 45,308 kits; 312 days at sea; £130,047. 15.03.1960: Off North Cape, Norway, (Sk J Boyle). Responded to a call for assistance from Hull trawler BANYERS (H255) (608grt/1940) (Sk S Digby) with crew fighting a fire in the stokehold which resulted in damage to engine room and crew accommodation. Connected and commenced tow. 16.03.1960: Delivered Honningsvåg. 17.04.1975: Landed. Laid up for disposal. 05.06.1975: Sold to Jacques Bakker & Zonen, Bruges, Belgium, for breaking up. 06.06.1975: Delivered. 31.07.1975: Hull registry closed - "sold to Belgians". 08.09.1975: Commenced breaking up.

CANADA	1368	237	100.7	C D Holmes	Alexandra Towing Co Ltd,
183812	23.02.1951		26.6	1000ihp 3-cyl	Liverpool
Steam tug	03.07.1951		12.0	12.0 knots	

Oil fired.   07.1951: Registered at Liverpool.   1969: Sold to Fratelli Barretta fu Domenico, Brindisi, Italy. Liverpool registry closed. Registered at Brindisi as **STREPITOSO**. 1976: Company restyled Fratelli Barretta, Brindisi. 1981: Company restyled Impresa Fratelli Barretta, Brindisi. 1985: Company restyled Impresa Fratelli Barretta fu Domenico e Giovanni Barretta Snc, Brindisi. 1986: Sold to Barretta (Fratelli) Snc, Brindisi. 1988: Company restyled Impresa Fratelli Barretta & Domenico e Giovanni Barretta Snc, Brindisi. 12.1988: Sold to Comifer Srl, Brindisi, for breaking up. 21.12.1988: Commenced demolition at Brindisi. 1989: Registry closed.

FORMBY	1369	237	100.7	C D Holmes	Alexandra Towing Co Ltd,
183813	09.03.1951		26.6	1000ihp 3-cyl	Liverpool
Steam tug	24.07.1951		12.0	12.0 knots	

Oil fired.   07.1951: Registered at Liverpool.   1969: Sold to Fratelli Barretta fu Domenico, Brindisi. Liverpool registry closed. Registered at Brindisi as **PODEROSO**.   1976: Company restyled Fratelli Barretta, Brindisi.   1981: Company restyled Impresa Fratelli Barretta, Brindisi. 1985: Company restyled Impresa Fratelli Barretta fu Domenico e Giovanni Barretta Snc, Brindisi.   1986: Sold to Barretta (Fratelli) Snc, Brindisi.   1988: Company restyled Impresa Fratelli Barretta & Domenico e Giovanni Barretta Snc, Brindisi.   07.2001: Removed from Lloyd's Register of Shipping - "Vessel's continued existence in doubt".

GLADSTONE	1370	237	100.7	C D Holmes	Alexandra Towing Co Ltd,
183816	23.04.1951		26.6	1000ihp 3-cyl	Liverpool
Steam tug	06.09.1951		12.0	12.0 knots	

Oil fired.   09.1951: Registered at Liverpool. 1968: Sold to Società Anonima Italiana Lavori Edili Marittimi (SAILEM), Palermo, Italy. Liverpool registry closed. Registered at Palermo as **ARCHIMEDE S**.   1997: Removed from Lloyd's Register of Shipping.

CAPE BEAVER	1371	396	140.1	Amos & Smith	National Sea Products Ltd,
194423	23.05.1951	133	26.7	605ihp 3-cyl	Halifax, Nova Scotia,
CFNo.41 Steam trawler	11.10.1951		12.8	11.5 knots	Canada

Oil fired. 10.1951: Registered at Halifax, NS (41).   17.10.1951: Sailed Selby for Halifax via Falmouth.   31.10.1951: Arrived Halifax. 1969: Sold to Ocean Fisheries Ltd, Halifax.   1971: Sold to J W Stephens, Halifax.   1971: Laying afloat at Halifax as a hulk. 06.1971: Halifax registry closed.   1974: Sold to The J P Porter Co Ltd, Halifax. To be rebuilt as a spoil hopper / sand carrier. 30.05.1974: Approved name change to J.P.P.602 but not taken up.   1977: J P Porter Co Ltd in liquidation. Sold to Nittolo Metals Inc, Cap St Ignace, Quebec, and broken up.

CAPE BRIER	1372	396	140.1	Amos & Smith	National Sea Products Ltd,
194427	21.06.1951	133	26.7	605ihp 3-cyl	Halifax, Nova Scotia,
CFNo.39 Steam trawler	21.11.1951		12.8	11.5 knots	Canada

Oil fired. 11.1951: Registered at Halifax, NS (39).   1969: Sold to Ocean Fisheries Ltd, Halifax.   1971: Sold to J W Stephens, Halifax. 1971: Noted as 'condemned' laying afloat at Halifax.   1974: Sold to The J P Porter Co Ltd, Halifax. Rebuilt as a spoil hopper/sand carrier and re-engined with twin 12-cylinder oil engines, 680bhp total, by General Motors Corp, Detroit. Registered at Halifax as **J.P.P.601**. 1977: J P Porter Co Ltd in liquidation.   08.1977: Laid up and offered for sale.   10.1980: Sold to Artlac Dredging Ltd, Notre-Dame-de-Pierreville, Ontario.   04.08.1980: Sailed Hamilton along with J.P.P.No.600 in tow of tug TUSKER for Quebec. In Quebec engines re-rated 1050bhp 8 knots. Halifax registry closed.   1980: Registered at Sorel as **KIM R.D**.   1983: Sold to Construction Canamount Inc, Montreal. 1989: Sold to CIBC Leasing Ltd, Montreal. Sorel registry closed. Registered at Montreal as **MARIE-LOU**. 1990: Leased to McKeil Work Boats, Hamilton, then to Dean Construction, Windsor, Ontario.   05.1992: At Hamilton, forecastle cut down to rails and bottom doors sealed. 22.06.1992: Sailed from Hamilton under tow along with J.P.P.600 bound Panama.   06.1992: Sold to Unida Transportadora S.A, Panama City, Panama. Montreal registry closed. Registered at Panama; 379grt, 241net.   2010: Sold to Arenera Balboa S.A. Panama City. 2015: Still in service.

RANA	1373	98	80.3	British Polar	Gaselee & Son Ltd,
184541	19.07.1951		21.6	750bhp 5-cyl	London
Motor tug	04.12.1951		8.6		

12.1951: Registered at London.   01.05.1965: Sold to Ship Towage (London) Ltd, London.   27.01.1969: Transferred to London Tugs Ltd, London.   01.01.1975: Sold to Alexandra Towing Co (London) Ltd, London.   03.1975: Transferred to Alexandra Towing Co Ltd, Swansea. 08.1979: Sold to Humphrey & Grey (Lighterage) Ltd, London. Registered at London as **REDRIFF**.   11.1982: Company wound up. 05.1983: Advertised for sale.   1984: Sold to Bennet Bros, London. Registered at London as **RANA**.   1990: Company restyled Medway Lighterage Ltd, London.   08.2006: Sold and converted to a houseboat, moored at Hoo, River Medway. No further details.

Arctic Warrior (1367)

(Author's collection)

A splendid view of **Formby** (1369) on trials.

(Cochrane archive)

*The **Canada** (1368) emerges from the Liverpool dock system and enters the River Mersey.*

(Stuart Emery collection)

*The **Gladstone** (1370) in Southampton Water and with original design of funnel.*

(Stuart Emery collection)

*This later image of **Gladstone**, again at Southampton, shows that the funnel has been shortened.*

(Stuart Emery collection)

The Canadian flag flies prominently at the stern of the **Cape Beaver** (1371) during her trials in the Humber.

(Cochrane archive)

The **Rana** (1373) at work in the Thames.

(M J Gaston, Stuart Emery collection)

MARGARET LOCKET 184562 Motor tug	1374 18.08.1951 03.01.1952	74	70.7 18.1 7.3	British Polar 460bhp 6-cyl 7.3 knots	Charrington Gardner Locket (London) Ltd, London

Launched as **MARGARET**. 01.1952: Registered at London as **MARGARET LOCKET**. 1965: Transferred to Charrington Lighterage Co Ltd., London. 1969: Sold to F T Everard & Sons Lighterage Ltd, Greenhithe. 1970: Registered at London as **P. B. EVERARD**. 1981: Sold to W G S Crouch & Sons Ltd., Greenhithe. Registered at London as **JAYNE SPEARING**. 1984: Sold to Tidal Marine & Dredging Services (Essex) Ltd, Grays, Essex. 1984: Sold to Riverway Development Co Ltd, London. 1988: Sold to J G Jakubait & Sons, London. 1992: Registry closed. Broken up.

GAVINA 185292 FD167 Motor trawler	1375 14.01.1952 05.05.1953	316 109	127.7 26.6 12.3	Mirrlees Bickerton & Day 772bhp 7-cyl 11.0 knots	J Marr & Son Ltd, Fleetwood

Cost to build £84,167. 06.05.1953: Registered at Fleetwood (FD167). 20.06.1959: Sold to Dinas Steam Fishing Co Ltd, Fleetwood. 07.1967: Sold to Peter & J Johnstone Ltd, Aberdeen, for £40,000. 20.07.1967: Fleetwood registry closed. 21.07.1967: Registered at Aberdeen (A871). 1968: Peter & J Johnstone Ltd acquired by J Marr & Son Ltd, Fleetwood. 26.08.1970: Sold to P W McLellan Ltd, Bo'ness, in an en bloc deal with LUNEDA (see yard no.1376). 09.1970: Commenced breaking up at Bo'ness. 29.10.1970: Aberdeen registry closed - "Vessel broken up".

LUNEDA 185293 FD175 Motor trawler	1376 13.02.1952 01.07.1953	316 109	127.7 26.6 12.3	Mirrlees Bickerton & Day 772bhp 7-cyl 11.0 knots	J Marr & Son Ltd, Fleetwood

Cost to build £84,275. 02.07.1953: Registered at Fleetwood (FD175). 12.1964: Sold to Peter & J Johnstone Ltd, Aberdeen. 30.12.1964: Fleetwood registry closed. 31.12.1964: Registered at Aberdeen (A754). 10.01.1968: Sold to Bon Accord Fish Selling Co Ltd, Aberdeen. 26.08.1970: Sold to P W McLellan, Bo'ness, in an en bloc deal with GAVINA (see yard no.1375). 09.1970: Commenced breaking up at Bo'ness. 29.10.1970: Aberdeen registry closed - "Vessel broken up".

CAPE ARGOS 194433 CFNo.35 Steam trawler	1377 31.10.1951 29.01.1952	399 135	140.1 26.7 12.8	Amos & Smith 605ihp 3-cyl 11.0 knots	National Sea Products Ltd, Halifax, Nova Scotia, Canada

Oil fired. 01.1952: Registered at Halifax, NS (35). Fishing out of Lunenberg. 1960s: Fishing out of Halifax. 23.03.1964: In thick fog stranded on George's Island. Refloated without damage. 20.02.1967: In thick fog stranded on George's Island. Refloated without damage. 11.1967: Sold to Metal Processors Ltd, Sydney, Nova Scotia, and broken up. 1967: Halifax registry closed.

CAPE BONNIE 194436 CFNo.32 Steam trawler	1378 15.11.1951 18.03.1952	399 135	140.1 26.7 12.8	Amos & Smith 605ihp 3-cyl 11.0 knots	National Sea Products Ltd, Halifax, Nova Scotia, Canada

Oil fired. 03.1952: Registered at Halifax, NS (32). 03.04.1952: Arrived Halifax. 21.02.1967: Inwards from the La Have Bank grounds in fog, snow and freezing spray (Sk Peter Hickey); eighteen crew in total. Stranded on the Sambro Ledges off Woody Island, one mile from the shore of Pennant Bay, 12 miles west of Halifax harbour. At about 4.00am sent out distress call "we are standing into immediate trouble". A few minutes later radioed, "aground and need immediate help". Crew abandoned but boats were dashed to pieces on the rocks and all eighteen lost. Halifax registry closed.

CAPE SAMBRO 195187 CFNo.27 Steam trawler	1379 27.03.1952 07.07.1952	399 135	140.1 26.7 12.8	Amos & Smith 605ihp 3-cyl 11.0 knots	National Sea Products Ltd, Halifax, Nova Scotia, Canada

Oil fired. 07.1952: Registered at Halifax, NS (27). 1966: Sold to Ocean Fisheries Ltd, Halifax. 1971: Laying afloat at Halifax as a hulk. 1974: Sold to The J P Porter Co Ltd, Halifax. Rebuilt as a spoil hopper/sand carrier and re-engined with twin 12-cylinder oil engines, 680bhp total, by General Motors Corp, Detroit; 8 knots. Registered at Halifax as **J.P.P.600**. Laid up at Halifax. 1977: The J P Porter Co Ltd in liquidation. 08.1977: Laid up and offered for sale. 08.1980: Sold to Artlac Dredging Ltd, Notre-Dame-de-Pierreville, Ontario. 04.08.1980: Sailed Hamilton along with J.P.P.No.601 in tow for Quebec. In Quebec engines re-rated to 1050bhp; 8 knots. Halifax registry closed. 1980: Registered at Sorel as **DAVID T.D.** 1983: Sold to Construction Canamount Inc, Montreal. 1989: Sold to CIBC Leasing Ltd, Montreal. Sorel registry closed. Registered at Montreal as **MARIE-SOL** (IMO 5061853). 1990: Leased to McKeil Work Boats, Hamilton, then to Dean Construction, Windsor, Ontario. 05.1992: At Hamilton, forecastle cut down to rails and bottom doors sealed. 05.06.1992: Sold to Dragarena S A, Panama City, Panama. Sorel registry closed. Registered at Panama, 379grt, 241net. 22.06.1992: Sailed from Hamilton under tow along with J.P.P.601 bound Panama. 2015: Still in service.

PRINCESS ELIZABETH 185133 H238 Motor trawler	1380 10.05.1952 16.10.1952	514 186	161.1 29.2 13.7	Crossley Bros 960bhp 8-cyl 13.2 knots	Saint Andrew's Steam Fishing Co Ltd, Hull

Contract price £154,800. 17.10.1952: Registered at Hull (H238). 21.10.1952: Sailed on first trip. 20.05.1958: Last landing at Hull, transferred to Grimsby. 12.06.1958: First landing at Grimsby. 19.10.1959: Sold to Southern Steam Trawling Co Pty Ltd, Port Adelaide, Australia. 29.11.1959: Registered at Hull as **SOUTHERN ENDEAVOUR** (H238). 08.06.1962: Hull registry closed. 06.1962: Registered at Port Adelaide. Engine re-rated to 1000bhp. 1970: Sold to Gulf Seafoods (Weipa) Pty Ltd, Brisbane. 1971: Sold to Blyth Greene Jourdain (Seafoods) Pty Ltd, Brisbane. 1974: Sold to Northern Seafoods Pty Ltd, Brisbane. 20.09.1979: Settled on bottom at Cattle Wharf, near Skelton Creek, Cairns. 14.04.1981: Present owner Mr B Murray was given two months to refloat vessel and remove. 18.06.1981: Legal proceedings against the owner began. 21.10.1981: Harbour Board sought quotations to raise and remove. 03.07.1982: Raised and disposed of by sinking in 16 fathoms some 65km from Cairns in position 16.39S 146.19E.

PRINCE CHARLES 185142 H249 Motor trawler	1381 09.07.1952 06.01.1953	514 184	161.1 29.2 13.7	Crossley Bros 960bhp 8-cyl 13.2 knots	Saint Andrew's Steam Fishing Co Ltd, Hull

Contract price £154,800.   28.01.1953: Registered at Hull (H249).   29.01.1953: Sailed on first trip.   1953: Made one trip to Greenland fishery.   31.12.1954: Last landing at Hull, transferred to Grimsby.   03.12.1955: Sailed Grimsby for Barents Sea grounds (Sk Thomas H Baskcomb); twenty-one crew.   23.12.1955: Homeward with Norwegian pilots onboard when some 60 miles west of Hammerfest in a snowstorm stranded on Karken Island. Norwegian motor vessel INGÖY (433grt/1950) rescued twelve crew from the shore but nine men including the skipper and a Norwegian pilot were lost. HNoMS NORDKYN (P.No.F309) and Hull trawler KINGSTON TOPAZ (H145) (794grt/1950) stood by overnight in the hope of finding more survivors.   1956: During the summer the wreck was refloated by Norwegian salvors and towed to Tromsø for inspection.   06.1956: Abandoned to underwriters.   13.06.1956: Sold to The British Trawlers Reinsurance Association Hull.   15.09.1956: Sold to Loch Fishing Co Ltd, Hull.   03.10.1956: Arrived Hull in tow of Norwegian salvage tug ULLER (319grt/1943).   Survey on slip revealed extensive damage to shell plating port side. Humber St. Andrew's Engineering Co Ltd, Hull, entrusted with repairs.   03.01.1957: Registered at Hull as **LOCH MELFORT** (H249).   11.05.1957: Sailed on first trip after repairs.   15.06.1957: New Certificate of Registration issued after being re-engined with 8-cylinder 1185bhp oil engine by Mirrlees, Bickerton & Day Ltd, Stockport; 12.5 knots and alteration of tonnage and other particulars, now 490grt, 170net.   03.1960: Outward for Norwegian coast grounds (Sk H Moon, Scarborough); twenty crew. In thick fog, in collision 65 miles off the mouth of the Humber with Indian motor vessel JALADHIR (6527grt/1957) sustaining heavy damage to fore end; no injuries to crew in either vessel. Returned to Hull stern first at 3 knots with JALADHIR in company until met by Hull tug FOREMAN (227grt/1959) which connected and assisted to dock. Repaired and returned to service.   11.08.1964: Last landing at Hull. Transferred within the Associated Fisheries Group to Wyre Trawlers Ltd, Fleetwood.   18.08.1964: Sold to Wyre Trawlers Ltd, Fleetwood.   15.09.1966: Registered at Fleetwood (FD228).   29.09.1966: Hull registry closed.   24.07.1968: Registered at Fleetwood as **WYRE CAPTAIN** (FD228).   01.07.1969: Became part of the British United Trawler (B.U.T.) fleet. Book value £45,806.   1975: Laid up.   06.06.1976: Sold to Mayer Newman & Co Ltd, London, and allocated to Porthleven Shipyard for breaking up.   10.06.1976: Arrived Porthleven in tow of motor tug OCEAN PULLER (145grt/1943).   1976: Fleetwood registry closed.

ST LEONARD 185289 FD179 Motor trawler	1382 21.08.1952 26.02.1953	275 101	115.5 25.1 11.1	Crossley Bros 600bhp 8-cyl 12.5 knots	Saint Andrew's Steam Fishing Co Ltd, Hull

27.02.1953: Registered at Fleetwood (FD179).   14.01.1955: Sold to Great Western Fishing Co Ltd, Aberdeen. Fishing out of Fleetwood.   09.1955: Transferred to fish out of Grimsby.   04.07.1956: Sold to W A Phillips, Anderson & Co Ltd, London.   01.1957: Transferred to fish out of Fleetwood.   08.02.1960: Sold to Fishery Products Ltd, St John's, Newfoundland.   09.02.1960: Sailed for Canada.   10.02.1960: Fleetwood registry closed.   02.1960: Registered at St John's as **ZEBRA**.   12.12.1966: Stranded on rocks one mile from Ile aux Morts, south-west coast of Newfoundland, and became a total loss. One crew lost.   1967: St John's registry closed.

ZARP 194588 Motor trawler	1383 20.09.1952 09.04.1953	276 100	115.5 25.1 11.1	Crossley Bros 600bhp 8-cyl 12.5 knots	Fishery Products Ltd, St John's, Newfoundland, Canada

04.1953: Registered at St John's.   1977: Non-ferrous metals and usable parts removed, hulk scuttled.   1977: St. John's registry closed.

GRIMSBY TOWN 184914 GY246 Steam trawler	1384 19.11.1952 09.07.1953	711 256	181.2 31.2 15.4	Amos & Smith 1250ihp 3-cyl 13.2 knots	Consolidated Fisheries Ltd Grimsby

Oil fired.   10.07.1953: Registered at Grimsby (GY246). Insured for £161,600.   20.06.1969: On Icelandic grounds, sister trawler HULL CITY (GY282) (see yard no.1385) disabled with trawl fouling propeller, connected and commenced tow. Delivered Ísafjörður.   20.03.1975: Sold to Mayer Newman & Co Ltd, London, and allocated to Fleetwood for breaking up.   Re-sold to Hammond Lane Foundry Co Ltd, Dublin.   26.06.1972: Grimsby registry closed.   07.1975: Arrived Dublin for breaking up.

HULL CITY 184916 GY282 Steam trawler	1385 17.01.1953 27.08.1953	711 256	181.2 31.2 15.4	Amos & Smith 1250ihp 3-cyl 13.2 knots	Consolidated Fisheries Ltd Grimsby

Oil fired.   28.08.1952: Registered at Grimsby (GY282). Insured for £161,600.   11.04.1954: On the Icelandic grounds (Sk John Searby).   In heavy weather attracted by a handkerchief waved on an oar, picked up eight Icelandic fishermen from a rubber dinghy after their Vestmannaeyjar motor fishing boat, GLADUR (VE270) (Sk þorleifur Guðjónsson), had capsized and foundered in a storm.   13/14.07.1960: Observed by auxiliary coastguard vessel ALBERT (201grt/1957) fishing off Langanes, north-east Iceland. Ordered to stop but when not heeded, warning shots fired. Radioed HMS UNDINE (P.No.F141) for help, warps cut and ran to seaward soon out-running the ALBERT.   01.1964: Homeward from the Icelandic grounds lost position and attempted to pass between Streymoy and Estyuroy, Faroe Islands, believing them to be Streymor and Vagar. Stranded on good bottom ground and came off on rising tide.   20.06.1969: Towed into Ísafjörður by sister trawler GRIMSBY TOWN (GY246) (see yard no.1384) with trawl fouling propeller. Cleared and proceeded to sea without a pilot, but stranded cutting the underwater telephone cable to the airfield. Refloated on rising tide with minimal damage.   07.04.1975: Sold to Mayer Newman & Co Ltd, London.   12.04.1975: Arrived Fleetwood for breaking up.   26.06.1975: Grimsby registry closed..

NORTHERN CROWN 184918 GY284 Steam trawler	1386 17.03.1953 26.11.1953	804 291	183.4 32.1 16.9	Amos & Smith 1150ihp 3-cyl 13.0 knots	Northern Trawlers Ltd, London

Oil fired.   11.1953: Registered at Grimsby (GY284).   06.10.1956: Sailed Grimsby for the fishing grounds on the west coast of Greenland (Sk Colin Newton); twenty crew.   10.10.1956: In view of reports of bad weather and poor fishing off Greenland, decided to fish off Iceland. At about 12.30 set course for Westmann Islands.   11.10.1956: At 06.50 with a south-westerly wind about Force 6, moderate to rough sea and swell with some rain at times, stranded in heavy seas on a reef close to Gant Rock, Eldey, about 8 miles south-west of Reykjanes Point, Iceland. With the engine room flooding and both lifeboats damaged, the crew took to the inflatable liferafts. The Icelandic gunboat THOR (693grt/1951) answered the distress call and guided to the area by an American aircraft arrived on the scene at 10.00 and took on board the whole crew as the trawler sank by the stern.   Grimsby registry closed.   12.01.1957: At formal Investigation, Sk Newton was adjudged to be at fault and his certificate was suspended for twelve months.

A splendid image of **Princess Elizabeth** (1380).

(Cochrane archive)

Grimsby Town (1384) dressed overall when new.

(Jonathan Grobler collection)

We keep football fans happy on both sides of the Humber with this fine image of **Hull City** (1385) in new condition.

(Cochrane archive)

NORTHERN SCEPTRE	1387	804	183.4	Amos & Smith	Northern Trawlers Ltd,
184919	17.03.1953	291	32.1	1100ihp 3-cyl	London
GY297 Steam trawler	26.11.1953		15.6	13.0 knots	

Oil fired. Cost to build £203,787. The last steam trawler out of Grimsby.
05.02.1954: Registered at Grimsby (GY297). 03.03.1954: Landed her first catch at Grimsby, 3014 kits, £8.821 gross (Sk H Self).
11.01.1967: On an Icelandic trip, north-west of Iceland grounded on an iceberg. Attended by Fleetwood trawler ROBERT HEWETT(LO65) which connected and pulled clear. With damaged steering gear, rudder and making water in the boiler room taken in tow by ROBERT HEWETT for Reykjavik. 13.01.1967: Delivered Reykjavik after a stormy two day passage. 01.07.1969: Became part of the British United Trawler (B.U.T.) fleet. B.U.T. livery adopted. Book value £62,750. 28.01.1975: Last landing. Laid up for disposal. 21.04.1975: Company restyled British United Trawlers Ltd. 03.03.1976: Sold to H Kitson Vickers (Engineers) & Sons Ltd. Sheffield for breaking up in an en bloc deal with NORTHERN JEWEL (see yard no.1366). 14.05.1976: Resold to London Demolition Co Ltd and allocated to Medway Secondary Metals Ltd, Rainham, Kent, for breaking up. 19.03.1979: Grimsby registry closed. 17.03.1979: Reported laid up at Bloors Wharf, Rainham, Kent.

LORD ALEXANDER	1388	790	183.0	C D Holmes	Lord Line Ltd,
185177 IMO 5211771	20.06.1953	277	32.2	1275ihp 3-cyl	Hull
H12 Steam trawler	07.04.1954		15.6	13.2 knots	

Oil fired. 02.11.1951: Contract signed for Associated Fisheries Trawling Co Ltd, Hull. Cost to build £187,980. Final cost, fitted out £195,831.
08.04.1954: Registered at Hull (H12) in ownership of Lord Line Ltd, Hull; last vessel built by Cochrane for this company.
03.05.1963: Associated Fisheries Group, Hull, re-organised. 01.10.1965: Transferred within Associated Fisheries Group to Hellyer Bros Ltd, Hull. Hellyer livery adopted. 24.05.1966: Sold to Hellyer Bros Ltd for £197,514. 01.07.1969: Became part of British United Trawlers (B.U.T.) when Ross Group amalgamated with Associated Fisheries. Book value £62,347. 12.03.1975: Landed. Laid up in St Andrew's Dock. 05.1975: Sold to Mayer Newman & Co Ltd, London. Towed to Santander. 09.05.1975: Sold to Desguaces y Salvamentos, Aviles, Spain. 26.05.1975: Arrived Aviles for breaking up. 05.05.1978: Hull registry closed - ".... sold to Spaniards".

WATERLOO	1389	200	92.6	C D Holmes	Alexandra Towing Co,
185484	24.10.1953		25.7	950ihp 3-cyl	Liverpool
Steam tug	27.04.1954		11.7	12.0 knots	

04.1954: Registered at Liverpool. 10.1962: Converted to burn oil fuel. 1972: Sold to Società Rimorchiatori Napoletani, Naples, Italy. 1972: Liverpool registry closed. Registered at Naples as **DRITTO**. 1989: Sold to Salvatore Palermo & Co for breaking up. 06.1989: Arrived Naples and broken up. 1989: Registry closed.

WALLASEY	1390	200	92.6	C D Holmes	Alexandra Towing Co,
185490	23.11.1953		25.7	950ihp 3-cyl	Liverpool
Steam tug	15.06.1954		11.7	12.0 knots	

06.1954: Registered at Liverpool. 1960s: Based Swansea. 09.1961: Converted to burn oil fuel. 1972: Sold to South Ocean Services Ltd, Portsmouth. 1972: Registered at Liverpool as **KENDIKEN**. 1970s: Laid up in Fareham Creek on various occasions. 04.1983: At Woolston, Southampton, laid up in poor condition. 1992: Removed from Lloyd's Register of Shipping.

CANNING	1391	200	92.6	C D Holmes	Alexandra Towing Co,
185492	05.02.1954		25.7	950ihp 3-cyl	Liverpool
Steam tug	29.06.1954		11.8	12.0 knots	

Oil fired. First oil-fired tug built for the company. 09.1954: Registered at Liverpool. 1966: Transferred to Swansea. 1974: Withdrawn from service. Last steam tug working in Bristol Channel ports. 08.1975: Sold to The Council of the City of Swansea for preservation. 2014: Preserved in Swansea. On National Historic Ships Register.

Not built	1392, 1393 Orders cancelled

FLEETWOOD LADY	1394	370	128.5	British Polar	Boston Deep Sea Fishing
185296	06.03.1954	126	26.7	700bhp 5-cyl	& Ice Co Ltd,
FD1 Motor trawler	09.09.1954		12.3	13.5 knots	Fleetwood

Built without aid of a WFA grant.
07.09.1954: Registered at Fleetwood (FD1). 30.11.1955: Company restyled Boston Deep Sea Fisheries Ltd, Fleetwood. 07.03.1956: Sailed Fleetwood on morning tide for fishing grounds (Sk Reginald Wright); fifteen crew. During afternoon anchored in Ramsey Bay to allow three Manx crewmen, the Lyall brothers and Allan Bradford, to visit families. At 4.00 pm. seven crew members left in the ship's boat for Ramsey. At 7.00 pm four crew returned in ship's boat leaving Sk. Reginald Wright, Allan Bradford and Eric Lyall ashore. At about 10.30pm. a small 13ft boat with six occupants left Ramsey to return onboard. At some time later the boat capsized and all occupants were thrown into the water. Alarm was raised after a body was seen floating just offshore and Ramsey lifeboat THOMAS CORBETT (Cox Cottier) was launched to search the area. Only bottom boards, identified as from the boat, were found.   All bodies were recovered from the shoreline except that of Sk Wright. 21.02.1962: Fleetwood registry closed. 01.03.1962: Sold to Acadia Fisheries Ltd, Mulgrave, Nova Scotia, Canada. 03.1962: Registered at St John's, Newfoundland, as **ACADIA KINGFISHER**. 21.10.1968: In heavy weather in Gulf of St Lawrence started to take in water and abandoned by crew to another company trawler. 22.10.1968: Company trawler connected and commenced tow but started to settle and foundered in approx position 48.50N 62.35W. 1968: St John's registry closed.

Northern Sceptre (1387)

Lord Alexander (1388)

A trio of tugs for the Alexandra Towing Co Ltd.

Waterloo (1389) at Swansea with a Blue Funnel cargo ship in the background.

(World Ship Photo Library, Stuart Emery collection)

Wallasey (1390) heading for the lock in Kings Dock, Swansea, and dressed overall for an unknown reason.

(World Ship Photo Library, Stuart Emery collection)

Canning (1391) enters Brocklebank Dock at Liverpool. She has been preserved at Swansea.

(Stuart Emery collection)

WROXHAM QUEEN 184000 LT53 Motor trawler	1395 05.04.1954 05.10.1954	179 60	102.0 22.0 10.9	Crossley Bros 350bhp 6-cyl 12.0 knots	Talisman Trawlers Ltd, West Hartlepool	

06.10.1954: Registered at Lowestoft (LT53). 06.04.1966: Sold to Talisman Trawlers (North Sea) Ltd, West Hartlepool. 11.09.1970: Sold to Henry J Lamprell, Buntingford, Herts, and Milford Haven. 05.08.1970: Registered at Lowestoft as **JADESTAR GYPSY** (LT53). Employed at times as an offshore platform standby safety vessel out of Lowestoft & Gt Yarmouth.. 05.10.1974: Sank at her moorings at Milford Haven, later raised and laid up. 1.09.1975: Sold by mortgagees to Keith & Mark Davies, Haverfordwest. 15.11.1976: Sold to Hubert Jones (Trawlers) Ltd, Swansea & Milford Haven. 1976: Towed motor trawler SALLY McCABE (YH88) (125grt/1030) from Milford Haven to Briton Ferry for breaking up. 13.04.1978: Hubert Jones (Trawlers) Ltd in liquidation. 02.06.1978: Sold at auction to Mr Peter Wright, Milford Haven, in an en bloc deal with BRENDA WILSON (LT80) (181grt/1954) and GEORGINA WILSON (HL10) (182grt/1955) for the sum £11,225. 1978: Sold to Thos W Ward Ltd, Sheffield, for breaking up at Briton Ferry. 01.04.1989: Lowestoft registry closed.

LUDHAM QUEEN 184002 LT87 Motor trawler	1396 03.05.1954 11.11.1954	179 60	102.0 22.0 10.9	Crossley Bros 350bhp 6-cyl 11.5 knots	Talisman Trawlers Ltd, West Hartlepool	

16.11.1954: Registered at Lowestoft (LT87). 06.04.1966: Sold to Talisman Trawlers (North Sea) Ltd, West Hartlepool. 12.06.1970: Sold to Henry J Lamprell, Buntingford, Herts, & Milford Haven. 25.09.1970: Registered at Lowestoft as **JADESTAR GLORY** (LT53). Employed at times as an offshore platform standby safety vessel out of Lowestoft & Gt Yarmouth. 16.01.1974: Off Arklow coast (Sk Harry John); five crew. Dodging in heavy weather, struck rocks off Roney Point in position 52.36N 06.10W. Arklow lifeboat attended but came afloat as they attempted to rig a breeches buoy and drifted down on Cahore Point. Crew of six took to liferafts and rescued by Arklow lifeboat. Vessel struck twice after going ashore, flooding her engine room and fish room. Subsequently moved further inshore and bottom plating set up and open to sea. 02.1974: Sold to Hammond Lane Industries Ltd, Dublin, 'as is'. 11.03.1974: Salvage vessel FOUR MILEY attended but was damaged, salvage crew taken off by helicopter. 11.1974: Salved, towed to Dublin and broken up. 29.03.1976: Lowestoft registry closed - "Vessel broken up".

OLIVEAN 184923 GY92 Motor trawler	1397 03.06.1954 14.12.1954	254 82	117.7 25.2 10.9	British Polar 640bhp 4-cyl 12.0 knots	Sir Thomas Robinson & Son (Grimsby) Ltd, Grimsby	

Cost to build £105, 000 (WFA Grant £25,000).
12.1954: Registered at Grimsby (GY92). 1976: Laid up. 08.1976: Company ceased trading. Repossessed by WFA on outstanding loan repayment. 10.1976: Sold to Dagon Fishing Co Ltd, Luton. 27.11.1976: Arrived Lowestoft. 02.12.1976: Grimsby registry closed. 03.12.1956: Registered at Lowestoft (LT392); did not fish from Lowestoft. 12.04.1977: Completed survey following conversion for offshore platform standby safety duties. 19.04.1977: Registered at Lowestoft. 27.04.1977: Registered at Lowestoft as **MUSTIQUE**. 16.12.1978: Arrived Falmouth with serious engine crankshaft damage. Towed to Lowestoft. 01.1979: Laid up at old Brooke Marine site, Oulton Broad. 28.10.1981: Registered at Lowestoft as **ROSS HERON** (see yard no.1468) to free the name. 29.10.1981: Sold to Henderson & Morez Ltd, Ilford, for breaking up at Gravesend. 21.01.1982: Sailed Lowestoft for River Thames in tow. 12.07.1984: Lowestoft registry closed - "Vessel broken up".

THESSALONIAN 184925 GY112 Motor trawler	1398 02.07.1954 25.01.1955	254 82	117.7 25.2 10.9	British Polar 640bhp 4-cyl 12.0 knots	Sir Thomas Robinson & Son (Grimsby) Ltd, Grimsby	

Built with aid of a WFA Grant.
15.12.1954: Registered at Grimsby (GY112). Insured for £127,200. 1976: Laid up. 08.1976: Company ceased trading. Repossessed by WFA on outstanding loan repayment. 19.10.1976: Sold to Dagon Fishing Co Ltd, Luton. 06.11.1976: Arrived Lowestoft. 03.12.1976: Registered at Lowestoft (LT272); did not fish from Lowestoft. 12.01.1977: Completed survey following conversion for offshore platform standby safety duties. 12.01.1977: Registered at Lowestoft. 26.05.1977: Registered at Lowestoft as **MARTINIQUE**. 12.1986: Sold to Cook Bros, New Holland, for breaking up. 22.01.1987: Sailed Lowestoft in company with TOBAGO (194grt/1960) also for breaking up. Arrived New Holland same day. Registry closed.

COLLENA 185298 FD20 Motor trawler	1399 14.09.1954 22.02.1955	334 113	127.5 26.5 13.0	British Polar 700bhp 5-cyl 13.0 knots	J Marr & Son Ltd, Fleetwood	

Contract at a cost of £125,000. Final cost £129,200 (WFA Grant £25,000).
23.02.1955: Registered at Fleetwood (FD20). 20.01.1964: Registered at Fleetwood as **VELIA** (FD20) when the name became free following sale of VELIA (FD116) (296grt/1951) to Italian owners. Change prompted due to radio traffic confusion with CORENA (see yard no.1439). 06.1971: Suffered major engine damage on the Icelandic grounds. Towed home and laid up. 03.1972: Sold to H Kitson Vickers & Sons (Engineers) Ltd, Sheffield. 08.1972: Commenced breaking up at Fleetwood. 18.09.1972: Fleetwood registry closed.

JOSEPH KNIBB 184926 GY2 Steam trawler	1400 16.11.1954 21.03.1955	442 154	139.7 28.2 13.5	Amos & Smith 800ihp 3-cyl 12.0 knots	Derwent Trawlers Ltd, Grimsby	

Oil fired. Cost to build £121,184 (WFA Grant £25,000)
22.03.1955: Registered at Grimsby (GY2). 01.1959: Sold to Ross Trawlers Ltd, Grimsby. 05.11.1960: Registered at Grimsby as **KENILWORTH** (GY2). 05.01.962: Registered at Grimsby as **ROSS KENILWORTH** (GY2). 04.05.1962: On an Icelandic trip (Sk Jack Simpson); sixteen crew. Some 20 miles off south-west coast of Iceland in gale force winds main injection box fractured, pumps could not cope with inflow of water. Grimsby trawler ROSS RODNEY (see yard no.1415) (Sk Jock Kerr) and Icelandic gunboat THOR (693grt/1951) responded to call for assistance. Eleven crew taken off by ROSS RODNEY and with pump from THOR onboard, connected and commenced tow. Pump could not cope and vessel started to settle and subsequently foundered 24 miles south of Malarrif, on the west coast of Iceland in approximate position 64.20N 24.10W. Remaining five crew taken off by THOR and eleven transferred from ROSS RODNEY; all landed at Reykjavik. 04.05.1962: Grimsby registry closed.

Ludham Queen *(1396)*

Collena *(1399)*

BERMUDA 186998 LT122　Motor trawler	1401 27.11.1954 28.04.1955	205 70	104.7 23.2 8.9	Ruston & Hornsby 403bhp　6-cyl 12.0 knots	Claridge Trawlers Ltd, Lowestoft

29.04.1955: Registered at Lowestoft (LT122).　First new build for Claridge owned companies.　04.05.1955: Arrived Lowestoft.　1956: Top Lowestoft trawler (Sk E Newbury) - £33,080 gross.　1972: Employed at times as an offshore platform standby safety vessel out of Lowestoft. 06.1980: Sold to Romford Steel Supplies Ltd, Romford, Essex.　29.06.1980: Sailed Lowestoft under tow for River Medway and broken up by Medway Secondary Metals Ltd, London, at Rainham, Kent.　02.10.1980: Lowestoft registry closed.

GRENADA 187000 LT130　Motor trawler	1402 13.12.1954 14.06.1955	205 70	104.7 23.2 8.9	Ruston & Hornsby 403bhp　6-cyl 12.0 knots	Dagon Fishing Co Ltd, Luton

15.06.1955: Registered at Lowestoft (LT130).　15.06.1955: Arrived Lowestoft.　18.06.1955: Sailed on first trip (Sk J W Howard). 09.01.1970: Stranded on Hopton beach, sustained rudder damage. Refloated at 22.25 by Lowestoft lifeboat and Yarmouth tug. 1973: Employed at times as an offshore platform standby safety vessel out of Lowestoft.　21.12.1973: Aground on Corton sands. Motor tug MARDYKE (37grt/1957) lost in going to her assistance.　04.08.1976: Completed survey following conversion for offshore platform standby safety duties.　09.11.1976: Lowestoft fishing registry closed.　23.05.1977: Registered at Lowestoft in the Registry of British Ships. 19.06.1986: Sold to G T Services, Barking, Essex, for breaking up.　21.08.1986: Sailed Lowestoft for River Thames.　24.11.1986: Lowestoft registry closed - "Vessel broken up".

BOSTON MONARCH 185301 FD19　Motor trawler	1403 27.01.1955 16.08.1955	466 164	137.5 28.0 13.75	British Polar 960bhp　6-cyl 12.0 knots	Saint Andrew's Steam Fishing Co Ltd, Hull

Built with aid of a WFA Grant.
16.08.1955: Registered at Fleetwood (FD19).　14.12.1965: Sold to Oceania S.p.A ,Trapani, Italy.　12.12.1965: Fleetwood registry closed. 12.1965: Registered at Trapani as **OCEANIA ROSA**.　1969: Sold to Achille Saltini, Trapani.　1972: Sold to Mankoadze Fisheries Ltd, Tema, Ghana.　1974: Sold to Achille Saltini, Trapani.　07.08.2009: Noted by Lloyd's Register of Shipping - "Vessel's continued existence in doubt".

JACINTA 185302 FD21　Motor trawler	1404 26.03.1955 06.09.1955	334 113	127.5 26.8 13.0	Mirrlees, Bickerton & Day 772bhp　7-cyl 11.25 knots	City Steam Fishing Co Ltd, Hull

Contract price £125,000. Built with aid of a WFA Grant.
06.09.1955: Registered at Fleetwood (FD21).　16.03.1962: Sold to Dinas Steam Trawling Co Ltd, Fleetwood.　06.07.1970: Sold to Bon Accord Steam Fishing Co Ltd, Aberdeen.　26.12.1971: Sold to P & W MacLellan Ltd, Bo'ness, and broken up.　26.12.1971: Fleetwood registry closed.

DORINDA 185305 FD22　Motor trawler	1405 07.05.1955 07.11.1955	334 113	127.5 26.8 13.0	British Polar 700bhp　5-cyl 11.25 knots	J Marr & Son Ltd, Fleetwood

Contract price £125,000. Built with aid of a WFA Grant.
08.11.1955: Registered at Fleetwood.　10.04.1973: Fleetwood fishing registry closed "no longer fishing".　12.11.1973: Sold to M R & T J Clarkson, Ampthill, Bedford.　21.07.1974: Sold to M R & T J Clarkson & T J McLaughlin, Ampthill, Bedford.　1975: Converted for offshore platform standby safety duties.　1978: Sold to Worldwide Surveys Ltd, Nassau, Bahamas, & Dallas, Texas. Fleetwood registry closed. Registered at Panama.　1998: Removed from Lloyd's Register of Shipping - "Vessel's continued existence in doubt".

WYRE VANGUARD 187841 FD36　Motor trawler	1406 22.06.1955 28.11.1955	338 119	127.5 27.0 13.5	Mirrlees, Bickerton & Day 736bhp　7-cyl 11.25 knots	Wyre Trawlers Ltd, Fleetwood

Cost to build £140,180 (WFA Grant £25,000).
15.12.1955: Registered at Fleetwood (FD36).　01.07.1969: Became part of British United Trawlers Ltd (B.U.T.) fleet.　1978: Transferred to fish out of Aberdeen.　1979: Laid up in Royal Albert Dock, East London.　1980: Sold to Court Bros Ltd, London.　Sold to R J Downes, Southend-on-Sea, for breaking up.　04.1980: Commenced breaking up at Southend.　15.09.1980: Fleetwood registry closed.

WYRE DEFENCE 187842 FD37　Motor trawler	1407 21.07.1955 11.01.1956	338 119	127.5 27.0 13.5	Mirrlees, Bickerton & Day 736bhp　7-cyl 11.25 knots	Wyre Trawlers Ltd, Fleetwood

Cost to build £138,372 (WFA Grant £25,000).
29.11.1955: Registered at Fleetwood (FD37).　01.07.1969: Became part of British United Trawlers Ltd (B.U.T.) fleet.　Book value £43,900. 1974: Converted for pair trawling (for herring) partnering WYRE MAJESTIC (see yard no.1414).　18.10.1974: After landing herring at Oban with WYRE MAJESTIC could not get an overnight berth and decided to sail for Fleetwood in company. Ahead through the Sound of Islay, WYRE MAJESTIC stranded at full speed at Rubh' a'Mhail, 4 miles north of Port Askaig, in position 55.53N 06.07W, badly holed and filling with water. The Port Askaig lifeboat attended, connected but with the trawler settling on the rocks, failed to move her. WYRE DEFENCE turned back, managed to connect but also failed as the tide ebbed. WYRE MAJESTIC was a total loss.　18.05.1978: Transferred to fish out of Aberdeen.　1979: Sold to J Gibson Johnson Ltd, Hull in an en bloc deal with WYRE REVENGE (see yard no.1413).　Sold to H Kitson Vickers & Sons (Engineering) Ltd, Sheffield.　12.1979: Sold to S Dalton Skip Hire Ltd, Edinburgh.　To be broken up at Bo'ness. 12.1979: In process of being broken up at Bo'ness.　14.01.1980: Fleetwood registry closed.

Bermuda *(1401) outward bound from Lowestoft.*

(Cochrane archive)

Wyre Defence *(1407)*

(Peter Horsley, author's collection)

KELLY 184936 GY6 Steam trawler	1408 19.09.1955 08.03.1956	448 154	137.5 28.0 14.25	Amos & Smith 750ihp 3-cyl 12.00 knots	Derwent Trawlers Ltd, Grimsby

Oil fired.   Cost to build £136,241 (WFA Grant £25,000).
07.03.1956: Registered at Grimsby (GY6).   29.01.1959: Sold to Ross Trawlers Ltd, Grimsby.   29.01.1959: Sold to Yorkshire Trawlers Ltd, Grimsby.   16.01.1962: Registered at Grimsby as **ROSS KELLY** (GY6).   10.04.1962: Sold to George F Sleight & Sons Ltd, Grimsby. 10.05.1966: Registered anew following alterations to particulars after re-engining by J & S Doig Ltd, Grimsby, and fitted with 6-cylinder 1260bhp oil engine by Ruston & Hornsby Ltd, Lincoln; lengthened by 23ft, cost £40,000. Remeasured to 160.5ft x 28.0ft; now 171net. Re-valued £206,003.   06.1967: Sold to Ross Steers Fisheries Ltd, St. John's, Newfoundland.   09.06.1967: Grimsby registry closed. Registered at St John's.   20.02.1968: Sold to George F Sleight & Sons Ltd, Grimsby.   03.1968: Returned to Grimsby.   St John's registry closed.   05.03.1968: Registered at Grimsby (GY125).   01.07.1969: Became part of British United Trawlers Ltd (B.U.T.) fleet. B.U.T. livery adopted. Book value £152,521.   10.1969: Sold to Goweroak Ltd, Grimsby.   03.1979: Sold to Ross Trawlers Ltd, Grimsby.   1981: Chartered by Colne Fishing Co Ltd, Lowestoft, and employed as an offshore platform standby safety vessel.   21.05.1982: Sold to George Craig & Sons Ltd, Aberdeen.   02.06.1982: Sold to The Colne Shipping Co Ltd, Luton.   12.07.1982: Grimsby registry closed.   07.1982: Registered at Lowestoft (LT125).   12.08.1982: Registered at Lowestoft as **CAICOS** (LT125) - PLN not applied to vessel.   1983: Converted to offshore platform standby safety vessel by Colne shore staff; bow thruster fitted by Brooke Marine Ltd, Oulton Broad.   03.05.1983: Fishing registry closed.   01.03.1984: Registered at Lowestoft as motor standby safety vessel.   09.1987: Sold to Liguria Maritime Ltd, Milton Regis, in an en bloc deal with NEVIS (see yard no.1431) and ST DAVIDS (576grt/1962).   29.09.1987: Sailed Lowestoft for River Medway towing ST DAVIDS.   08.10.1987: Delivered Milton Creek and broken up.   30.03.1988: Lowestoft registry closed - "Vessel broken up".

ROSS LION 184938 GY216 Motor trawler	1409 17.10.1955 18.04.1956	274 89	115.0 25.0 12.0	Ruston & Hornsby 605bhp 6-cyl 11.0 knots	Derwent Trawlers Ltd, Grimsby

Cost to build: £114,975 (WFA Grant £25,000).   First trawler built for Ross Group with an oil engine.
17.04.1956: Registered at Grimsby (GY216).   29.01.1959: Sold to Ross Trawlers Ltd, Grimsby.   29.10.1959: Sold to Yorkshire Trawlers Ltd, Grimsby.   10.04.1962: Sold to George F Sleight & Sons Ltd, Grimsby.   02.1963: Sold to Ross Lake Canada Ltd, St. John's, Newfoundland. 13.02.1963: Grimsby registry closed.   02.1963: Registered at St. John's, Newfoundland.   1973: Sold to Burgeo Fish Industries Ltd, Burgeo, Newfoundland.   1988: Removed from Lloyd's Register of Shipping - "continued existence in doubt"..

DINAS 187844 FD55 Motor trawler	1410 16.12.1955 30.05.1956	439 155	137.5 28.0 13.75	Mirrlees, Bickerton & Day 1012bhp 8-cyl 11.0 knots	Dinas Steam Trawling Co Ltd, Fleetwood

Built with aid of a WFA Grant.
31.05.1956: Registered at Fleetwood (FD55).   Insured for £146,000.   30.05.1959: Company taken over by J Marr & Son Ltd, Fleetwood. 06.1959: Transferred to fish out of Hull.   20.06.1959: Sailed for Icelandic grounds.   04.07.1959: At Hull landed 2,312 kits, £5,141 gross. 14.02.1967: At Hull landed from an Icelandic trip 632 kits, £3,525 gross.   Transferred to fish out of Fleetwood.   24.04.1971: Sold to J Marr & Son Ltd, Fleetwood.   17.09.1971: When outwards for the fishing grounds gutted by fire and towed back to Fleetwood; five crewmen died in the fire. Three crewmen were later convicted of arson and jailed for 5 years.   11.1971: Returned to service after refit on the Clyde. 13.04.1973: Transferred to fish out of Aberdeen.   15.06.1973: Sold to J Marr (Aberdeen) Ltd, Aberdeen.   21.01.1974: Fleetwood fishing registry closed - "no longer fishing".   Employed on offshore platform standby safety duties.   04.1976: Sold to Albert Draper & Son Ltd, Hull. 12.08.1976: Breaking up commenced at the Victoria Dock slipway.   Registry closed.

Not built	1411 Order cancelled

WYRE MARINER 187848 FD34 Steam trawler	1412 13.02.1956 23.07.1956	656 244	169.0 30.0 16.0	Amos & Smith 900ihp 3-cyl 14.00 knots	Wyre Trawlers Ltd, Fleetwood

Oil fired. Cost to build £188,894.
23.07.1956: Registered at Fleetwood (FD34).   10.04.1968: Transferred within Associated Fisheries Group to Northern Trawlers Ltd, London. 04.1968: Fleetwood registry closed.   26.04.1968: Registered at Grimsby (GY2).   05.08.1968: Registered at Grimsby as **NORTHERN SUN** (GY2).   01.07.1969: Became part of British United Trawlers Ltd (B.U.T.) fleet. B.U.T. livery adopted. Book value £75,500.   20.11.1974: Last landing. Laid up for disposal.   21.04.1975: Company restyled British United Trawlers (Grimsby) Ltd.   28.06.1976: Sold to Blyth Shipbreakers & Repairers Ltd, Blyth, for breaking up at Blyth.   28.05.1976: Grimsby registry closed.   06.1976: Breaking commenced.

WYRE REVENGE 187850 FD432 Motor trawler	1413 28.03.1956 18.09.1956	338 119	127.5 27.0 13.5	Mirrlees, Bickerton & Day 736bhp 7-cyl 11.25 knots	Wyre Trawlers Ltd, Fleetwood

Cost to build £148,081 (WFA Grant £30,000).
18.09.1956: Registered at Fleetwood (FD432).   01.07.1969: Became part of British United Trawlers Ltd (B.U.T.) fleet.   Book value £45,800. 07.04.1978: Transferred to fish out of Aberdeen.   1979: Sold to J Gibson Johnson Ltd, Hull, in an en bloc deal with WYRE DEFENCE (see yard no.1407).   Sold to H Kitson Vickers & Sons (Engineering) Ltd, Sheffield.   12.1979: Sold to S Dalton Skip Hire Ltd, Edinburgh. 12.1979: In process of breaking up at Bo'ness.   14.01.1980: Fleetwood registry closed.

Kelly (1408)

(Cochrane archive)

Ross Lion (1409)

(Cochrane archive)

Wyre Mariner (1412)

(Cochrane archive)

WYRE MAJESTIC	1414	338	127.5	Mirrlees, Bickerton & Day	Wyre Trawlers Ltd,
187853	12.05.1956	119	27.0	736bhp 7-cyl	Fleetwood
FD433 Motor trawler	12.11.1956		13.5	11.25 knots	

Cost to build £151,549 (WFA Grant £30,000).
12.11.1956: Registered at Fleetwood (FD433).   01.07.1969: Became part of British United Trawlers Ltd (B.U.T.) fleet. Book value £45,800.
1974: Converted for pair trawling (for herring) partnering WYRE DEFENCE (see yard no.1407).   18.10.1974: After landing herring at Oban with WYRE DEFENCE, could not get an overnight berth and Sk Derek Reader decided to sail for Fleetwood in company with WYRE DEFENCE. Following WYRE DEFENCE through the Sound of Islay, with the Bosun on watch, stranded at full speed at Rubh' a'Mhail, 4 miles north of Port Askaig, in position 55.53N 06.07W; badly holed and filling with water. The Port Askaig lifeboat connected but with the trawler settling on the rocks, failed to move her. WYRE DEFENCE, which had turned back, then connected but also failed as the tide ebbed. The lifeboat took off five crew. Sk Reader, Phil Huff (Mate) and Charlie Timmins (Chief Engineer) stayed on board to assist the tug due to arrive the following day   19.10.1974: Around high water tug connected but this attempt failed and all further attempts also failed.
01.11.1974: Declared a constructive total loss.   17.12.1974: Wreck sold "as lying" to Leven Divers, Kinlochleven, who stripped usable and non-ferrous items.   20.12.1974: Fleetwood registry closed.   Sk Reader was subsequently found at fault for not being in the wheelhouse and his ticket was suspended.   The Bosun, John Pirie, admitted to having had a lot to drink.

RODNEY	1415	751	182.5	Amos & Smith	Derwent Trawlers Ltd,
184952	10.07.1956	283	32.0	1350ihp 3-cyl	Grimsby
GY34 Steam trawler	17.01.1957		17.0	13.00 knots	

Oil fired. Cost to build £233,500. Built without WFA Grant. First trawler to dispense with lifeboats and carry inflatable life-rafts and a work boat under a Schat davit.
18.01.1957: Registered at Grimsby (GY34).   29.01.1959: Sold to Ross Trawlers Ltd, Grimsby.   29.10.1959: Sold to Yorkshire Trawlers Ltd, Grimsby.   10.04.1962: Sold to George F Sleight & Sons Ltd, Grimsby.   15.01.1962: Registered at Grimsby as ROSS RODNEY (GY34).
04.05.1962: On an Icelandic trip (Sk Jock Kerr) when some 20 miles off SW coast of Iceland in gale force winds with Icelandic gunboat THOR (693grt/1951), went to the assistance of ROSS KENILWORTH (see yard no.1400), unable to cope with inflow of water after main injection box fractured. Eleven crew taken off and with pump from THOR onboard, connected and commenced tow. Pump could not cope and vessel started to settle and subsequently foundered 24 miles south of Malarrif, on the west coast of Iceland in approximate position 64.20N 24.10W.   Remaining five crew taken off by THOR and eleven transferred; all landed at Reykjavik.   30.09.1967: Sold to Ross Trawlers Ltd, Grimsby.   03.04.1968: Registered anew following alteration of particulars and re-engining with 8-cylinder 1825bhp oil engine by Ruston & Hornsby Ltd, Lincoln. Now 697grt, 246net. Re-valued £364,145.   01.07.1969: Became part of British United Trawlers Ltd (B.U.T.) fleet. B.U.T. livery adopted. Book value £202,143.   1980: Sold to Jason Ford Group of Companies.   03.1980: Arrived Hartlepool for breaking up.   12.05.1981: Grimsby registry closed.

ROSS TIGER	1416	355	127.5	Ruston & Hornsby	Derwent Trawlers Ltd,
184953	22.09.1956	127	26.5	840bhp 7-cyl	Grimsby
GY398 Motor trawler	26.02.1957		13.0	11.5 knots	

Cost to build £140,473 (WFA Grant £30,000)
27.07.1957: Registered at Grimsby (GY398).   29.01.1959: Sold to Ross Trawlers Ltd, Grimsby.   29.10.1959: Sold to Yorkshire Trawlers Ltd, Grimsby.   10.04.1962: Sold to George F Sleight & Sons Ltd, Grimsby.   01.07.1969: Became part of British United Trawlers Ltd (B.U.T.) fleet. B.U.T. livery adopted. Book value £46,767.   10.09.1969: Sold to Goweroak Ltd, Grimsby.   30.09.1978: Sold to George F Sleight & Sons Ltd, Grimsby.   29.09.1982: Sold to British United Trawlers Ltd Grimsby.   26.07.1985: Sold to Cam Shipping Ltd, Grimsby.
19.11.1985: Registered at Grimsby as CAM TIGER following conversion to offshore platform standby safety vessel by George Prior Engineering Ltd, Lowestoft.   1992: Donated to Great Grimsby Borough Council for use as a museum ship, renamed ROSS TIGER and berthed at the Grimsby National Heritage Centre.

KIPLING	1417	448	137.5	Amos & Smith	Derwent Trawlers Ltd,
184955	22.10.1956	154	28.0	800ihp 3-cyl	Grimsby
GY38 Steam trawler	15.04.1957		14.25	11.5 knots	

Cost to build £153,466 (WFA Grant £30,000)
06.04.1957: Registered at Grimsby (GY38).   29.01.1959: Sold to Ross Trawlers Ltd, Grimsby.   29.10.1959: Sold to Yorkshire Trawlers Ltd Grimsby.   10.04.1962: Sold to George F Sleight & Sons Ltd, Grimsby.   01.01.1962: Registered at Grimsby as ROSS KIPLING (GY38).
24.08.1966: Registered anew following alterations to particulars, lengthening and re-engining by Ross Engineering Ltd, Grimsby, at cost of £40,000. Fitted with a 6-cylinder 1440bhp oil engine by Ruston & Hornsby Ltd, Lincoln. Remeasured to 160.6 feet; 469grt, 160net. Re-valued £201,984.   07.1967: Sold to Ross Steers Fisheries Ltd, St John's, Newfoundland.   28.07.1967: Grimsby registry closed. Registered at St John's.   03.1968: Sold to George F Sleight & Sons Ltd, Grimsby. St John's registry closed.   27.03.1968: Registered at Grimsby (GY126).   01.07.1969: Became part of British United Trawlers Ltd (B.U.T.) fleet. B.U.T. livery adopted. Book value £153,644.
30.09.1969: Sold to Goweroak Ltd, Grimsby.   30.09.1979: Sold to Ross Trawlers Ltd, Grimsby.   21.05.1982: Sold to George Craig & Sons Ltd, Aberdeen.   27.07.1982: Grimsby registry closed.   08.1982: Registered at Aberdeen as GRAMPIAN FREEDOM (A126) following conversion to offshore platform standby safety vessel.   1991: Sold to Cook Bros, New Holland, Lincolnshire, and broken up.   Aberdeen registry closed.

SAMARIAN	1418	331	127.6	British Polar	Onward Steam Fishing
181382	20.11.1956	118	26.6	950bhp 5-cyl	Co Ltd,
GY445 Motor trawler	17.05.1957		11.9	12.0 knots	Grimsby

Built with aid of a WFA Grant.
18.06.1957: Registered at Grimsby (GY445). Insured for £155,800.   14.04.1963: Fishing off Iceland (Sk Peter Newby) in very heavy weather hit by a huge wave, lost all radio and radar contact and boat deck swept clean. HMS MALCOLM (P.No.F88) on fishery protection duties in the vicinity and unable to raise her on radio, mistakenly reported her capsized and lost with all hands. Escorted into Akureyri by Hull trawler CAPE CAMPBELL (H383).   16.08.1965: Off Rora Head, Hoy, in collision with Aberdeen motor trawler LORWOOD (A400) (237grt/1960). Both trawlers put into Scrabster to effect temporary repairs.   14.10.1976: Sold to Colne Fishing Co Ltd, Luton.
21.10.1976: Arrived Lowestoft.   30.11.1976: Grimsby registry closed.   03.12.1976: Registered at Lowestoft (LT545).   12.07.1977: Surveyed after conversion to offshore platform standby safety vessel.   03.11.1977: Off the Shetlands in approximate position 60.48N 01.27E developed serious engine problems and was towed home, and laid up on return.   13.08.1979: Sold to David Potts (Steel) Ltd, Grimsby, and sailed Lowestoft under tow for Grimsby.   10.09.1979: Broken up at Doig's Slipway, Grimsby.   Grimsby registry closed.

Wyre Revenge (1413)
landing at Hull.

(Author's collection)

Wyre Majestic
(1414).

(Peter Horsley,
author's collection)

Kipling (1417)

(Jonathan Grobler
collection)

RHODESIAN 181385 GY457　Motor trawler	1419 03.01.1957 25.07.1957	331 118	127.6 26.6 11.9	British Polar 950bhp　5-cyl 12.0 knots	Onward Steam Fishing Co Ltd, Grimsby

Built with aid of a WFA Grant.
26.07.1957: Registered at Grimsby (GY457). Insured for £156,400.　22.12.1967: Remeasured to 331grt, 110net.　14.10.1976: Sold to Taylor Steam Fishing Co Ltd, Grimsby.　22.12.1976: Registered at Grimsby as **SANDO** (GY457).　1979: Sold to Konidaris Brothers S.A, Piræus, Greece.　08.10.1979: Grimsby registry closed.　10.1979: Registered at Piræus as **NAFTILOS**.　1982: Sold to Martechnic Marine Services Ltd, Piræus.　1986: Sold to Alikos Shipping Co, Piræus.　Registered at Piræus as **THEODOROS**.　01.2001: Removed from Lloyd's Register of Shipping - "Vessel's continued existence in doubt".

Not built	1420, 1421　Orders cancelled				

KASHMIR 181388 GY43　Steam trawler	1422 18.02.1957 12.09.1957	448 154	137.5 28.0 14.25	Amos & Smith 800ihp　3-cyl 12.00 knots	Derwent Trawlers Ltd, Grimsby

Oil fired.　Cost to build: £159,430 (WFA Grant £30,000).
13.09.1957: Registered at Grimsby (GY43).　29.01.1959: Company restyled Ross Trawlers Ltd, Grimsby.　29.10.1959: Sold to Yorkshire Trawlers Ltd, Grimsby.　10.01.1962: Registered at Grimsby as **ROSS KASHMIR** (GY43).　10.04.1962: Sold to George F Sleight & Sons Ltd, Grimsby.　30.09.1967: Sold to Ross Trawlers Ltd, Grimsby.　05.05.1967: Registered anew following alterations to particulars after re-engining and lengthening by J S Doig Ltd, Grimsby, at a cost of £40,000. Fitted with 6-cylinder 1440bhp oil engine by Ruston & Hornby Ltd, Lincoln. Lengthened to 160.6ft; 489grt, 171net. Re-valued £192,036.　07.12.1967: Sold to Ross Steers Fisheries Ltd, St. Johns, Newfoundland.　23.01.1968: Grimsby registry closed. Registered at St John's.　02.1968: St John's registry closed.
28.02.1968: Registered at Grimsby (GY122).　20.03.1968: Sold to George F Sleight & Sons Ltd, Grimsby.　30.09.1969: Sold to Goweroak Ltd, Grimsby.　01.07.1969: Became part of British United Trawlers fleet (B.U.T.). B.U.T. livery adopted. Book value £108,714.　09.1978: Sold to Ross Trawlers Ltd, Grimsby.　21.05.1982: Sold George Craig & Sons Ltd, Aberdeen.　24.05.1982: Grimsby registry closed.
05.1982: Registered at Aberdeen as **GRAMPIAN FAME** (A122). Converted to offshore platform standby safety vessel.　1987: Sold to Phoenix Stichting, Amsterdam (part of the Greenpeace Group).　1988-89: Converted at Hamburg to a schooner-rigged research vessel. Re-engined with twin oil engines, 1000kW/1340bhp total, by Deutz-MWM, Mannheim; 10 knots cruising speed; 555grt.　11.1989: Registered at Amsterdam as **RAINBOW WARRIOR**.　16.08.2011: Decommissioned.　08.2011: Sold to Friendship (NGO), Dhaka, Bangladesh, which provides essential basic services to the most inaccessible and hard-to-reach areas of the country.　29.08.2011: Arrived Chittagong for conversion to a hospital ship on completion; registered at Chittagong as **RONGDHONU** (IMO 5300481), which means rainbow in Bangla. 2015: Still in service.

ROSS LEOPARD 181391 GY491　Motor trawler	1423 01.05.1957 21.10.1957	355 127	127.5 26.5 13.0	Ruston & Hornsby 840bhp　7-cyl 11.5 knots	Derwent Trawlers Ltd, Grimsby

Cost to build £142,515 (WFA Grant £30,000).
22.10.1957: Registered at Grimsby (GY491).　22.01.1959: Company restyled Ross Trawlers Ltd, Grimsby.　29.10.1959: Sold to Yorkshire Trawlers Ltd, Grimsby.　10.04.1962: Sold to George F Sleight & Sons Ltd, Grimsby.　01.07.1969: Became part of the British United Trawlers (B.U.T.) fleet. B.U.T. livery adopted. Book value £52,163.　30.09.1969: Sold to Goweroak Ltd, Grimsby.　29.09.1982: Sold to British United Trawlers (Grimsby) Ltd, Grimsby.　01.1985: Sold to Cam Shipping Ltd, Hull.　03.04.1985: Converted to offshore platform standby safety vessel by George Prior Engineering Ltd, Lowestoft.　11.07.1985: Registered at Grimsby as **CAM LEOPARD**.　1992: Laid up following new regulations for offshore platform standby safety vessels.　05.1992: Sold to Andrew Standring, Jeremy Clinton & Toby Flitton, London, as Ross Leopard Ltd. Converted to a nightclub and entertainment centre moored on the Albert Embankment, Lambeth. Renamed **ROSS LEOPARD**.　09.2002: Towed by motor tug WARRIOR (58grt/1937) to a temporary mooring just above Battersea railway bridge.　01.2006: Renamed **LEOPARD**.　07.2008: Laid up at Ostend.　11.10.2010: Sold to Van Heyghen Recycling S.A., Ghent, for breaking up.

ROSS JAGUAR 181394 GY494　Motor trawler	1424 31.05.1957 10.12.1957	355 127	127.5 26.5 13.0	Ruston & Hornsby 840bhp　7-cyl 11.5 knots	Derwent Trawlers Ltd, Grimsby

Cost to build: £144,977 (WFA Grant £30,000).
10.12.1957: Completed acceptance trials in Humber estuary.　11.12.1957: Registered at Grimsby (GY494).　12.12.1957: Sailed Grimsby on first trip to North Sea grounds (Sk J Sheader).　21.12.1957: At Grimsby landed 150 kits, £834 gross.　22.01.1959: Company restyled Ross Trawlers Ltd, Grimsby.　29.10.1959: Sold to Yorkshire Trawlers Ltd, Grimsby.　10.04.1962: Sold to George F Sleight & Sons Ltd, Grimsby.　01.07.1969: Became part of the British United Trawlers (B.U.T.) fleet. B.U.T. livery adopted. Book value £53,863.　30.09.1969: Sold to Goweroak Ltd, Grimsby.　30.09.1978: Sold to George F Sleight & Sons Ltd, Grimsby.　29.09.1982: Sold to British United Trawlers (Grimsby) Ltd, Grimsby.　07.08.1985: Sold to Cam Shipping Ltd, Hull.　1985: Converted to offshore platform standby safety vessel by George Prior Engineering Ltd, Lowestoft.　16.12.1985: Registered at Grimsby as **CAM JAGUAR**　06.1992: Laid up following new regulations for offshore platform standby safety vessels.　Sold to Ron & Helen Devereux, Australia. Converted at Grimsby to a 3-masted schooner (sail area 345 square metres) with passenger accommodation. Remeasured to 414gt, 124net. Registered at Grimsby as **JAGUAR** (IMO 5300467). 2014: Offered for sale.

KHARTOUM 181396 GY47　Motor trawler	1425 29.06.1957 23.01.1958	422 147	137.5 28.0 14.0	Ruston & Hornsby 1170bhp　6-cyl 12.5 knots	Derwent Trawlers Ltd, Grimsby

Cost to build £168,261 (WFA Grant £30,000).
24.01.1958: Registered at Grimsby (GY47).　29.01.1959: Company restyled Ross Trawlers Ltd Grimsby.　29.10.1959: Sold to Yorkshire Trawlers Ltd, Grimsby.　10.01.1962: Registered at Grimsby as **ROSS KHARTOUM** (GY47).　10.04.1962: Sold to George F Sleight & Sons Ltd, Grimsby.　03.05.1963: Registered anew following alteration to particulars after lengthening by Drypool Engineering & Dry Dock Co Ltd at Hull, cost £40,000. Now 160.6 ft; 507grt, 197net. Re-valued £212,855.　31.03.1966: Sold to Ross Trawlers Ltd, Grimsby.　08.1967: Sold to Ross Steers Fisheries Ltd, St John's, Newfoundland.　24.08.1967: Grimsby registry closed. Registered at St John's.　20.02.1968: Sold to Ross Trawlers Ltd, Grimsby.　03.1968: St.John's registry closed.　11.03.1968: Re-registered at Grimsby (GY120).　01.07.1969: Became part of the British United Trawlers (B.U.T.) fleet. B.U.T. livery adopted. Book value £87,535. 13.09.1969: Sold to Goweroak Ltd, Grimsby.　02.1978: Converted to offshore platform standby safety vessel.　30.09.1978: Sold to Ross Trawlers Ltd Grimsby.　19.12.1980: Stranded 8 miles north of Aberdeen after engine failed in heavy weather. Eight crew rescued by helicopter and vessel abandoned as a total loss. 17.02.1981: Sold to T Youngston Ltd, Peterhead, and broken up in situ.　09.07.1982: Grimsby registry closed.

KANDAHAR 181399 GY506 Motor trawler	1426 11.09.1957 28.02.1958	422 147	137.5 28.0 13.5	Ruston & Hornsby 1170bhp 6-cyl 12.0 knots	Derwent Trawlers Ltd, Grimsby

Cost to build: £169,646 (WFA Grant £30,000)
28.02.1958: Registered at Grimsby (GY506). 29.01.1959: Owners restyled Ross Trawlers Ltd, Grimsby. 29.10.1959: Sold to Yorkshire Trawlers Ltd, Grimsby. 01.01.1962: Registered at Grimsby as **ROSS KANDAHAR** (GY506). 10.04.1962: Sold to G F Sleight & Sons Ltd, Grimsby. 03.05.1963: Registered anew following alteration to particulars. Lengthened by Drypool Engineering & Dry Dock Co Ltd at Hull at a cost of £40,000. Remeasured to 160.6 ft; 507grt, 197net. Re-valued £203,241. 31.03.1966: Sold to Ross Trawlers Ltd, Grimsby. 29.08.1967: Sold to Queen Steam Fishing Co Ltd, Grimsby. 09.1967: Sold to Ross Steers Fisheries Ltd, St. John's, Newfoundland. 21.09.1967: Grimsby registry closed. Registered at St.John's, Newfoundland. 20.02.1968: Sold to Queen Steam Fishing Co Ltd, Grimsby. 03.1968: St John's registry closed. 11.03.1968: Registered at Grimsby (GY120). 01.07.1969: Became part of the British United Trawlers (B.U.T.) fleet. B.U.T. livery adopted. Book value £116,010. 13.09.1969: Sold to Goweroak Ltd, Grimsby. 1976: Laid up. 06.1978: Sold to H Kitson Vickers & Sons (Engineering) Ltd, Sheffield, for breaking up. 19.12.1979: Grimsby registry closed.

ROSS PANTHER 181401 GY519 Motor trawler	1427 25.10.1957 17.04.1958	355 127	127.5 26.5 13.0	Ruston & Hornsby 840bhp 7-cyl 12.0 knots	Derwent Trawlers Ltd, Grimsby

Cost to build £151,993 (WFA Grant £30,000).
18.04.1958: Registered at Grimsby (GY519). 29.01.1959: Company restyled Ross Trawlers Ltd, Grimsby. 29.10.1959: Sold to Yorkshire Trawlers Ltd Grimsby. 10.04.1962: Sold to George F Sleight & Sons Ltd, Grimsby. 20.09.1967: Sold to Ross Trawlers Ltd, Grimsby. 01.07.1969: Became part of the British United Trawlers (B.U.T.) fleet. B.U.T. livery adopted. Book value £56,731. 30.09.1969: Sold to Goweroak Ltd, Grimsby. 30.09.1978: Sold to George F Sleight & Sons Ltd, Grimsby. 29.09.1982: Sold to British United Trawlers, Grimsby. 10.05.1985: Sold to Cam Shipping Ltd, Hull. Converted to offshore platform standby safety vessel by Globe Engineering Ltd, Hull. 31.07.1985: Registered at Grimsby as **CAM PANTHER**. 18.08.1987: Laid up for disposal as unsuitable. 04.06.1992: Sold to H Webster & Sons. 12.1992: Commenced breaking up at Grimsby. 1993: Grimsby registry closed.

ROSS COUGAR 181403 GY531 Motor trawler	1428 09.12.1957 02.06.1958	355 127	127.5 26.5 13.0	Ruston & Hornsby 840bhp 7-cyl 12.0 knots	Derwent Trawlers Ltd, Grimsby

Cost to build £151,954 (WFA Grant £30,000).
03.06.1958: Registered at Grimsby (GY531). 29.01.1959: Company restyled Ross Trawlers Ltd, Grimsby. 29.10.1959: Sold to Yorkshire Trawlers Ltd, Grimsby. 10.04.1962: Sold to George F Sleight & Sons Ltd, Grimsby. 30.09.1967: Sold to Ross Trawlers Ltd, Grimsby. 01.07.1969: Became part of the British United Trawlers (B.U.T.) fleet. B.U.T. livery adopted. Book value £58,106. 30.09.1969: Sold to Goweroak Ltd, Grimsby. 30.09.1978: Sold to George F Sleight & Sons Ltd, Grimsby. 29.09.1982: Sold to British United Trawlers (Grimsby) Ltd, Grimsby. 07.1985: The last side fishing trawler to land at Grimsby. 03.09.1985: Sold to Cam Shipping Co, Hull. Converted to offshore platform standby safety vessel by Globe Engineering Ltd, Hull. 11.06.1986: Registered at Grimsby as **CAM COUGAR**. 1989: Re-engined with a 6-cylinder 605bhp oil engine by Ruston & Hornsby Ltd, Lincoln, 11 knots. 06.1992: Laid up following new regulations for offshore platform standby safety vessels. 03.1993: Sold to Mohamad Abdelsalam Ossman, Tartous, Syria, and registered to Fayza Shipping Co Ltd, Limassol, Cyprus. 10.1992: Grimsby registry closed. 10.1992: Registered at Tartous as **OMAR** (IMO 5300352). 2015: Believed still in service.

EVERTON 168602 GY58 Steam trawler	1429 02.02.1958 03.09.1958	884 323	187.6 33.2 16.3	Amos & Smith 1450ihp 3-cyl 13.5 knots	Consolidated Fisheries Ltd, Grimsby

Oil fired. Cost to build £241,600. By gross registered tonnage the biggest side fishing trawler built for Grimsby owners.
04.09.1958: Registered at Grimsby (GY58). 1971: New boiler fitted at Newcastle. 26.05.1973: During the 2nd "Icelandic Cod War" (Sk George Mussel) strayed from the Royal Navy protected zone and stopped by gunfire from the Icelandic Coast Guard vessel ÆGIR (927grt/1968) with eight non-explosive rounds fired. The rounds penetrated the shell plating on the port side, some of them below the waterline, and at one stage the trawler settled low in the water. Repairs carried out by the British tug STATESMAN I (1167grt/1965) and the frigate HMS JUPITER (P.No.F60). No one was injured in the incident. 21.04.1976: Sold to Mayer Newman & Co Ltd, London, for £27,000. 10.11.1976: Arrived Sittingbourne, Kent, and broken up. 14.12.1976: Grimsby registry closed.

Not built	1430 Order cancelled

KELVIN 168603 GY60 Steam trawler	1431 21.04.1958 30.09.1958	448 164	137.5 28.0 14.25	Amos & Smith 800ihp 3-cyl 13.0 knots	Derwent Trawlers Ltd, Grimsby

Last oil fired steam trawler built for a Grimsby owner. Cost to build £166,684 (WFA Grant £30,000). To new design with transom stern.
01.10.1958: Registered at Grimsby (GY60). 09.01.1959: Company restyled Ross Trawlers Ltd, Grimsby. 10.1959: Sold to Yorkshire Trawlers Ltd, Grimsby. 01.1962: Registered at Grimsby as **ROSS KELVIN** (GY60). 04.1962: Sold to George F Sleight & Sons Ltd, Grimsby. 11.1966: Lengthened and re-engined by J S Doig Ltd, Grimsby. Fitted with 6-cylinder 1260bhp oil engine by Ruston & Hornsby Ltd, Lincoln. Remeasured to 160.5ft; 468grt, 159net. Re-valued £192,000. 10.1967: Sold to Ross Trawlers Ltd, Grimsby. 03.1968: Sold to George F Sleight & Sons Ltd, Grimsby. 09.1968: Sold to Ross Trawlers Ltd, Grimsby. 01.07.1969: Became part of the British United Trawlers (B.U.T.) fleet. B.U.T. livery adopted. Book value £105,210. 30.09.1969: Sold to Goweroak Ltd, Grimsby. 1981: Converted to offshore platform standby safety vessel. 1981: Chartered to The Colne Shipping Co Ltd, Luton. 21.05.1982: Sold to George Craig & Sons Ltd, Aberdeen. 02.06.1982: Sold to The Colne Shipping Co Ltd, Luton. 12.07.1982: Grimsby registry closed. 18.08.1982: Registered at Lowestoft as **NEVIS**. 28.09.1987: Sold to Liguria Maritime Ltd, Milton Regis, in an en bloc deal with CAICOS (see yard no.1408) and ST DAVIDS (576grt/1962). 28.09.1987: Sailed Lowestoft for River Medway. 07.10.1987: Delivered to Sittingbourne and broken up. 30.03.1988: Lowestoft registry closed - "Vessel broken up".

Rhodesian (1419)

(Cochrane archive)

Ross Cougar (1428)

(Cochrane archive)

Kelvin (1431) leaving Humber Dock, Hull, on trials.

(Cochrane archive)

OSAKO	1432	325	128.0	Mirrlees	Diamonds Steam Fishing
168604	20.05.1958	122	26.0	736bhp 7-cyl	Co Ltd,
GY600 Motor trawler	12.11.1958		11.5	13.0 knots	Grimsby

Built with the aid of a WFA Grant.
13.11.1958: Registered at Grimsby. Insured for £140,600. 21.12.1967: Remeasured to 324grt, 109net. 01.1981: Laid up, surveys overdue. 18.01.1982: Grimsby registry closed - "sold to Kuwait subjects". Was to have been sold en bloc with OGANO (see yard no.1442) to Yousif R H Qabazard, Kuwait, but sale was not completed. 1982: Sold to George Crutwell & George Shotton, Tyneside, for breaking up. 1983: Sold to CPA Salvage Ltd, Bromley, Kent. Engaged in recovering electrical cable in the Solent. 04.01.1984: Further converted for salvage work. 22.11.1984: Sold to Offshore Marine Salvage Ltd, Guernsey. 04.1985: Sold by Order of the Admiralty Marshal at Ipswich to Joe O'Connor, Plymouth, on behalf of Anglo Spanish interests. Restored to a fishing vessel. 22.01.1986: Sold to Lasin Ltd, Cork, Irish Republic. Registered at Cork (C70). 26.02.1991: Foundered whilst outward bound from Vigo, Spain, towards fishing grounds in approximate position 44.38N 09.36W. All crew rescued. 03.1991: Cork registry closed.

YESSO	1433	325	128.1	Mirrlees	Japan Fishing Co Ltd,
168607	03.07.1958	122	26.7	736bhp 7-cyl	Grimsby
GY610 Motor trawler	09.12.1958		11.5	13.0 knots	

Built with aid of a WFA Grant.
10.12.1958: Registered at Grimsby (GY610). Insured for £140,600. 21.12.1967: Remeasured to 324grt, 109net. 21.03.1981: Sold to Cleanseas Oil Pollution Control Ltd, Windsor, Berkshire. 22.6.1981: Arrived Grangemouth in tow of tug EUGENIO (31grt/1957). Converted to a pollution control vessel by Grangemouth Dockyard Co Ltd. 09.10.1981: Grimsby registry closed. 10.1981: Registered at Grangemouth as CLEANSEAS I. 1984: Sold to In The Footsteps of Scott 1984 Antarctic Expedition. After hull strengthening and arcticisation, registered at Grangemouth as SOUTHERN QUEST. 03.11.1984: Sailed from UK for Antarctica. 01.1986: While proceeding through the Ross Sea towards McMurdo Sound to collect three British explorers, became fast in the ice, 4 miles from Beaufort Island in approximate position 76.57S 167.13E. 11.01.1986: The working of the pack ice caused the fore end to crush and overheated machinery caught fire in the engine room. While the crew was dealing with the fire, the shell plating amidships, under considerable pressure, was breached and freezing water entered the engine room. Crew abandoned on to the ice just before midnight and ship foundered stern first shortly after. Twenty-four crew and passengers rescued by helicopter from the American Antarctic expedition base. Registry closed.

ANGUILLA	1434	228	102.5	Ruston & Hornsby	Clan Steam Fishing Co
301551 IMO 5017747	16.08.1958	82	23.5	403bhp 6-cyl	(Grimsby) Ltd,
LT67 Motor trawler	16.01.1959		11.5	12.0 knots	Lowestoft

27.01.1959: Registered at Lowestoft (LT67). 31.03.1972: Sold to Claridge Trawlers Ltd, Lowestoft. 04.12.1975: Lowestoft fishing registry closed. 07.09.1976: Surveyed after conversion to offshore platform standby safety vessel. 02.11.1986: Sold to Southard Trawlers Ltd, Milford Haven c/o Milford Haven Fish Merchants Association. 18.12.1986: Converted back to fishing. Registered at Lowestoft 2nd class fishing vessel (LT67). 01.04.1989: Lowestoft fishing registry closed. 02.04.1990: Lowestoft registry closed. 05.1991: Laid up. 07.1991: Company in receivership. 03.1992: Sold by the Receivers, Cork Gully, to Supreme Fishing Co Ltd, Belfast, c/o Westcoasting BV, IJmuiden. Subsequently sold to Rekker Warenhuis, Netherlands, along with ANTIGUA (204grt/1957) for conversion for private use as sailing vessel. 1997: Sold to Smitsco Ltd, Franeker, Amsterdam. 01.2007: Registered at Franeker as ANQUILLA. 04.2007: Remeasured to 186gt, 55net. 2015: Believed still in service.

MONTSERRAT	1435	228	102.5	Ruston & Hornsby	Huxley Fishing Co Ltd,
301512	29.09.1958	82	23.5	403bhp 6-cyl	Lowestoft
LT64 Motor trawler	20.02.1959		11.5	12.0 knots	

23.02.1959: Registered at Lowestoft (LT64). 24.05.1976. Lowestoft fishing registry closed. 24.08.1976: Converted to offshore platform standby safety vessel. 06.1987: Sold to Pesca Fisheries Ltd, Milford Haven, (Anglo Spanish principals) in an en bloc deal with CELITA (131grt/1957). 12.06.1987: Registered at Lowestoft as a 2nd class fishing vessel (LT376). 07.1987: Sold to Victor Montenegro Feijoo, Vigo, Spain. 30.07.1987: Sailed Lowestoft towing GUANA (228grt/1959). 08.1987: Arrived Vigo for breaking up. 23.04.1989: Lowestoft registry closed.

ROSS JACKAL	1436	355	127.5	Ruston & Hornsby	Ross Trawlers Ltd,
168611	09.12.1957	127	26.5	840bhp 7-cyl	Grimsby
GY637 Motor trawler	02.06.1958		13.0	12.0 knots	

Cost to build £152,214 (WFA Grant £30,000).
09.04.1959: Registered at Grimsby (GY637). 29.10.1959: Sold to Yorkshire Trawlers Ltd, Grimsby. 10.04.1962: Sold to George F Sleight & Sons Ltd, Grimsby. 30.09.1967: Sold to Ross Trawlers Ltd, Grimsby. 01.07.1969: Became part of the British United Trawlers (B.U.T.) fleet. B.U.T. livery adopted. Book value £65,678. 30.09.1969: Sold to Goweroak Ltd, Grimsby. 29.09.1982: Sold to British United Trawlers (Grimsby) Ltd, Grimsby. 20.08.1985: Sold to Cam Shipping Co, Hull. Converted to offshore platform standby safety vessel by George Prior Engineering Ltd, Lowestoft. 04.09.1985: Registered at Grimsby as CAM JACKAL. 23.12.1987: Sold to P M Vieira, Lisbon, Portugal. 24.09.1992: Sold to Pescasada S C, Sada, La Coruña, Spain. Converted to fish for sharks in Angolan waters. 05.05.1993: Laid up at Ferrol. 01.11.1993: Grimsby registry closed. 15.11.1993: Sold to Desguaces de la arena, La Coruña, Spain, for breaking up. 29.11.1993: Completed breaking up.

Not built	1437 Order cancelled

Osako (1432)

(Cochrane archive)

Yesso (1433)

(Cochrane archive)

Ross Jackal (1436)

(Cochrane archive)

NAVENA	1438	353	128.5	Mirrlees, Bickerton & Day	J Marr & Son Ltd,
187869	12.12.1958	125	27.0	875bhp 7-cyl	Fleetwood
FD172 Motor trawler	21.05.1959		13.5	CPP - 12.5 knots	

Cost to build £173,700. Built with aid of a WFA Grant. First trawler in Marr ownership fitted with a controllable pitch propeller. First Fleetwood trawler with a transom stern.
21.05.1959: Registered at Fleetwood. 28.01.1969: Transferred to Peter & J Johnson Ltd, Aberdeen, to fish out of Aberdeen.
02.01.1973: Sold to J Marr (Aberdeen) Ltd, Aberdeen. 06.12.1973: In a winter gale and seeking shelter on the north side of Copinsay, stranded in the early hours on a reef 600 yds offshore at the north end under the cliffs. At 0617 her distress call was answered by several vessels including BOUNTIFUL (K67) (49grt), COASTAL EMPRESS (A455) (250grt/1960), GLEN AFFRIC (A175) (114grt/1971) and the Kirkwall lifeboat, but impossible to approach due to high winds and heavy swell. The Deerness LSA team assembled at Kirkwall airport ready to be lifted by helicopter to Copinsay. The crew remained in the wheelhouse and the Kirkwall lifeboat reported that the trawler was in a critical position, only the bridge deck remained above water. At 1012 all twelve crew were rescued by helicopter (British Airways S61) from Dyce which was on standby at Kirkwall, landing them initially at Deerness before taking them to Kirkwall. The wreck is in approximate position 58.54N 02.40W. 13.12.1973: Fleetwood registry closed.

CORENA	1439	353	128.5	Mirrlees, Bickerton & Day	J Marr & Son Ltd,
301876	27.01.1959	125	27.0	736bhp 7-cyl	Fleetwood
FD173 Motor trawler	23.06.1959		13.5	CPP - 12.0 knots	

Cost to build £188,700. Built with aid of a WFA Grant.
20.06.1959: Registered at Fleetwood (FD173). 24.07.1970: Sold to Ranger Fishing Co Ltd, North Shields. 04.10.1971: Sold to Forward Motor Trawlers Ltd, Aberdeen. 30.01.1973: Fishing registry closed. 1976: Converted to offshore platform standby safety vessel.
22.03.1978: Sold to Colne Fishing Co Ltd, Luton. 06.1978: Registered at Lowestoft as **TRINIDAD**. 02.03.1982: Company re-styled The Colne Shipping Co Ltd. 01.1987: Sold to Liguria Maritime Ltd, Sittingbourne, Kent. Mid-01.1987: Sailed Lowestoft for River Medway and broken up at Milton Creek. 20.07.1987: Registry closed.

GALILÆAN	1440	328	127.6	British Polar	Sir Thomas Robinson & Son
168616	11.03.1959	102	26.6	950bhp 5-cyl	(Grimsby) Ltd,
GY603 Motor trawler	01.09.1959		11.7	12.0 knots	Grimsby

Built with aid of a WFA Grant.
26.08.1959: Registered at Grimsby (GY603). Insured for £164,300. 17.05.1968: Remeasured to 328grt, 101net. 20.09.1975: Sold to Taylor Trawlers Ltd, Grimsby. 21.12.1975: Registered at Grimsby as **NANAO** (GY603). 02.1981: Sold en bloc with KYOTO (see yard no.1441) to Curtis Consolidated Ltd, Dominican Republic. 09.02.1981: Grimsby registry closed. 02.1981: Registered at Panama.
1997: Removed from Lloyd's Register of Shipping - "Vessel's continued existence in doubt".

EPHESIAN	1441	328	127.6	British Polar	Sir Thomas Robinson & Son
301816	10.04.1959	102	26.6	950bhp 5-cyl	(Grimsby) Ltd,
GY604 Motor trawler	05.10.1959		11.7	12.0 knots	Grimsby

Built with aid of a WFA Grant.
01.10.1959: Registered at Grimsby (GY604). Insured for £164,300. 16.05.1968: Remeasured to 328grt, 101net. 16.10.1975: Sold to Taylor Trawlers Ltd, Grimsby. 18.12.1975: Registered at Grimsby as **KYOTO** (GY604). 02.1981: Sold en bloc with NANAO (see yard no.1440) to Curtis Consolidated Ltd on behalf of Eastern Equates Ltd, Dominican Republic. 09.02.1981: Grimsby registry closed.
02.1981: Registered at Panama. 1997: Removed from Lloyd's Register of Shipping - "Vessel's continued existence in doubt".

OGANO	1442	320	125.0	Mirrlees	Taylor Steam Fishing Co Ltd,
301817	23.05.1959	103	26.6	736bhp 7-cyl	Grimsby
GY608 Motor trawler	14.10.1959		11.5	12.0 knots	

Built with aid of a WFA Grant.
14.10.1959: Registered at Grimsby (GY608). Insured for £154,200. 21.12.1967: Remeasured to 320grt, 102net. 18.01.1982: Grimsby registry closed - "sold to Kuwait subjects". Was to have been sold en bloc with OSAKO (see yard no.1432) to Yousif R H Qabazard, Kuwait but sale was not completed. 1982: Sold to George Crutwell & George Shotton, Tyneside, for breaking up. 1983: Sold to Joe O'Connor, Plymouth, representing Anglo Spanish interests. Returned to fishing out of Vigo, Spain. 1986: Sold to Anglo Spanish interests represented by De-Tec Securities Ltd, Plymouth. Registered at Sligo (SO779). 03.1989: Laid up at Plymouth with engine damage. Sold by Order of the Admiralty Marshal to S Evans & Sons Ltd, Garston, for breaking up. 1989: Sold to W Bustard, London. Re-engined at Garston and fitted with a 898bhp/670kW oil engine by Mirrlees Blackstone Ltd, Stockport. 01.1990: Sold to Wedgeford Associates Ltd, Donegal, Co Donegal. Returned to fishing. 1990s: Laid up in Killybegs, Co Donegal. 1997: Scuttled off St John's Point, Dunkineely, Co Donegal, as a dive site; lies in 37 - 50m. Sligo registry closed.

ROSS CHEETAH	1443	354	127.5	Ruston & Hornsby	Yorkshire Trawlers Ltd,
301819	22.06.1959	116	26.5	705bhp 7-cyl	Grimsby
GY614 Motor trawler	24.11.1959		13.0	11.0 knots	

Cost to build £151,180 (WFA Grant £37,500).
25.11.1959: Registered at Grimsby (GY614). 10.04.1962: Sold to George F Sleight & Sons Ltd, Grimsby. 01.07.1969: Became part of the British United Trawlers (B.U.T.) fleet. B.U.T. livery adopted. Book value £64,175. 30.09.1967: Sold to Ross Trawlers Ltd, Grimsby.
30.09.1969: Sold to Goweroak Ltd, Grimsby. 29.09.1982: Sold to British United Trawlers (Grimsby) Ltd, Grimsby. 27.03.1985: Sold to Cam Shipping Co Ltd, Grimsby. 20.05.1985: New certificate issued following survey after conversion to offshore platform standby safety vessel by George Prior Engineering Ltd, Lowestoft. 29.05.1985: Registered at Grimsby as **CAM CHEETAH**. 06.1992: Sold Pescasada S C, Sada, La Coruña, Spain. 18.06.1993: Grimsby registry closed. Converted back to a trawler; on completion registered at Panama.
08.1998: Sold to Quasar Shipping Corp, Panama. Panama registry closed. Registered at São Tomé, São Tomé & Príncipe, Gulf of Guinea.
12.2000: Registered at São Tomé as **RIO CONGO**. 01. 2001: Sold to undisclosed owners registered in Lomé, Togo. 05.2010: Sold to Recuperaciones Siderurgicas, Santander, Spain, and broken up. Registry closed.

ROSS LYNX 301823 GY626 Motor trawler	1444 22.07.1959 01.02.1960	354 116	127.5 26.5 13.0	Ruston & Hornsby 705bhp 7-cyl 11.0 knots	Ross Trawlers Ltd, Grimsby

Cost to build £151,180 (WFA Grant £37,500).
27.01.1960: Registered at Grimsby (GY626).   01.07.1969: Became part of the British United Trawlers (B.U.T.) fleet. B.U.T. livery adopted. Book value £65,855.   30.09.1969: Sold to Goweroak Ltd, Grimsby.   23.03.1972: When some 10 miles north of Holborn Head reported man missing presumed lost overboard. Began search and joined by Thurso lifeboat THE THREE SISTERS but nothing found.   29.09.1982: Sold to British United Trawlers (Grimsby) Ltd, Grimsby.   20.03.1985: At Grimsby landed last trip.   27.03.1985: Sold to Cam Shipping Co Ltd, Hull. 10.06.1985: New certificate issued following survey after conversion to offshore platform standby safety vessel by George Prior Engineering Ltd, Lowestoft.   26.06.1985: Registered at Grimsby as **CAM LYNX**.   06.1992: Laid up following new regulations for offshore platform standby safety vessels.   11.1993: Sold to Bideawhile One Hundred and Twenty Nine Ltd, Ipswich.   07.1994: Sold to Takaoka Shipping Corp, Panama.   07.1994: Grimsby registry closed.   03.2001: Sold to undisclosed owners.   2014: Vessel's continued existence in doubt.

Not built	1445, 1446 Orders cancelled				

ROSS PUMA 301838 GY646 Motor trawler	1447 15.03.1960 30.08.1960	352 116	127.5 26.5 13.0	Ruston & Hornsby 1080bhp 6-cyl 13.2 knots	Ross Trawlers Ltd, Grimsby

Cost to build £152,525 (WFA Grant £37,500).
29.08.1960: Registered at Grimsby (GY646).   01.04.1968: In a NW gale and with heavy seas running, in the early hours, ran hard aground on the Little Rackwick shoals beneath the cliffs, two miles NNW of Tor Ness, Hoy, Orkney Islands. At 0123 a distress call was received by Wick radio. Trawlers ROSS RENOWN (see yard no.1462) and WILLIAM WILBERFORCE (H200) (698grt/1959) responded but could not approach. Longhope lifeboat T.G.B. (Cox Daniel Kirkpatrick) was launched and in very difficult conditions picked up the 15 crewmen from the ship's liferafts. (Cox Kirkpatrick was awarded the Silver Third-Service Clasp by the RNLI). Driven further ashore by gale force winds and a heavy swell the wreck broke up in position 58.48.1N 03. 20.5W.   29.04.1969: Grimsby registry closed - "Total loss".

PHILADELPHIAN 301830 GY636 Motor trawler	1448 02.11.1959 02.06.1960	272 88	119.5 25.1 11.4	British Polar 790bhp 5-cyl 11.5 knots	Sir Thomas Robinson & Son (Grimsby) Ltd, Grimsby

Built with the aid of a WFA Grant.
30.05.1960: Registered at Grimsby (GY636). Insured for £169,600.   16.05.1968: Remeasured to 272grt, 86net.   10.1976: Sir Thomas Robinson & Son (Grimsby) Ltd ceased trading. Laid up.   13.01.1977: Sold to A L Richards & Co Ltd, Grimsby.   03.1977: Sold to R Claeys, Ostende, Belgium.   23.03.1977: Grimsby registry closed. Converted to a beam trawler.   1978: Renamed **BELGIUM FISHERMAN** (not recorded in Belgian Register).   1979: Re-engined with a 6-cylinder 1447bhp oil engine by Anglo-Belgium Co N.V., Gent. CP propeller. 1980: Sold to Noordvisserij N.V, Zeebrugge.   08.04.1980: Registered at Zeebrugge as **PHILADELPHIAN** (Z319). Remeasured to 264gt, 75net.   1984: Sold to Robert Dewaele, Zeebrugge.   16.05.1986: Sold to Robert Dewaele-Baert, Zeebrugge.   21.10.1988: Sold to NV Dirk, Zeebrugge.   1989: Sold to The Gambia.   20.11.1989: Zeebrugge registry closed.   06.1991: Removed from Lloyd's Register of Shipping - "Vessel's continued existence in doubt".

JUDÆAN 301837 GY644 Motor trawler	1449 02.12.1959 25.07.1960	272 86	117.0 25.0 11.4	British Polar 790bhp 5-cyl 11.5 knots	Sir Thomas Robinson & Son (Grimsby) Ltd, Grimsby

Built with the aid of a WFA Grant.
15.07.1960: Registered at Grimsby (GY644). Insured for £169,600.   16.05.1968: Remeasured to 272grt, 86net.   10.1976: Sir Thomas Robinson & Son (Grimsby) Ltd ceased trading.   18.10.1976: Sold to Dagon Fishing Co Ltd, Luton.   30.11.1976: Grimsby registry closed. Registered at Lowestoft (LT125). Converted to offshore platform standby safety vessel.   06.1977: Registered at Lowestoft as **ABACO**. 17.07.1984: Caught fire in North Sea, in approximate position 53.52N 01.59E and abandoned by crew who were picked up by trawler BLACKBURN ROVERS (GY706) (408grt/1962). The following day with fire extinguished, towed back to Lowestoft, declared a constructive total loss with extensive fire damage to superstructure. Laid up.   1985: Sold to Henderson & Morez, Gravesend, for breaking up. 16.11.1985: Sailed Lowestoft in tow of tug EUGENIO (31grt/1957) for River Thames.   17.11.1985: Delivered Gravesend.   Registry closed.

ROSS GENET 301841 GY650 Motor trawler	1450 30.05.1960 05.10.1960	352 116	127.5 26.5 13.0	Ruston & Hornsby 1025bhp 6-cyl 11.0 knots	Ross Trawlers Ltd, Grimsby

Cost to build £153,630 (WFA Grant £37,500).
05.10.1960: Registered at Grimsby (GY650).   30.09.1961: Sold to Dennis Roberts Fishing Co Ltd, Grimsby.   31.03.1966: Sold to Ross Trawlers Ltd, Grimsby.   01.07.1969: Became part of the British United Trawlers (B.U.T.) fleet. B.U.T. livery adopted.   Book value £72,056. 30.09.1969: Sold to Goweroak Ltd Grimsby.   29.09.1982: Sold to British United Trawlers (Grimsby) Ltd Grimsby.   27.02.1985: Sold to Cam Shipping Co Ltd, Hull. Converted to offshore platform standby safety vessel by Globe Engineering Ltd, Hull.   09.05.1985: Registered at Grimsby as **CAM GENET**.   31.12.1985: Remeasured to 337grt, 114net.   06.1992: Laid up following new regulations for offshore platform standby safety vessels.   10.1993: Sold to Takaoka Shipping Corp, Panama.   10.1993: Grimsby registry closed.   1994: Converted back to a side trawler.   05.1999: Registered at São Tomé, São Tomé & Principe, Gulf of Guinea.   01.2001: Registered at São Tomé as **CENTAUR**. 07.2005: Sold to Malonia Trading SA, Panama.   São Tomé registry closed.   Registered at Panama as **CENTAURO 1**.   08.2008: Noted in Lloyd's Register of Shipping as a total loss.

Navena *(1438)*

(Cochrane archive)

Judaean *(1449)*

(Cochrane archive)

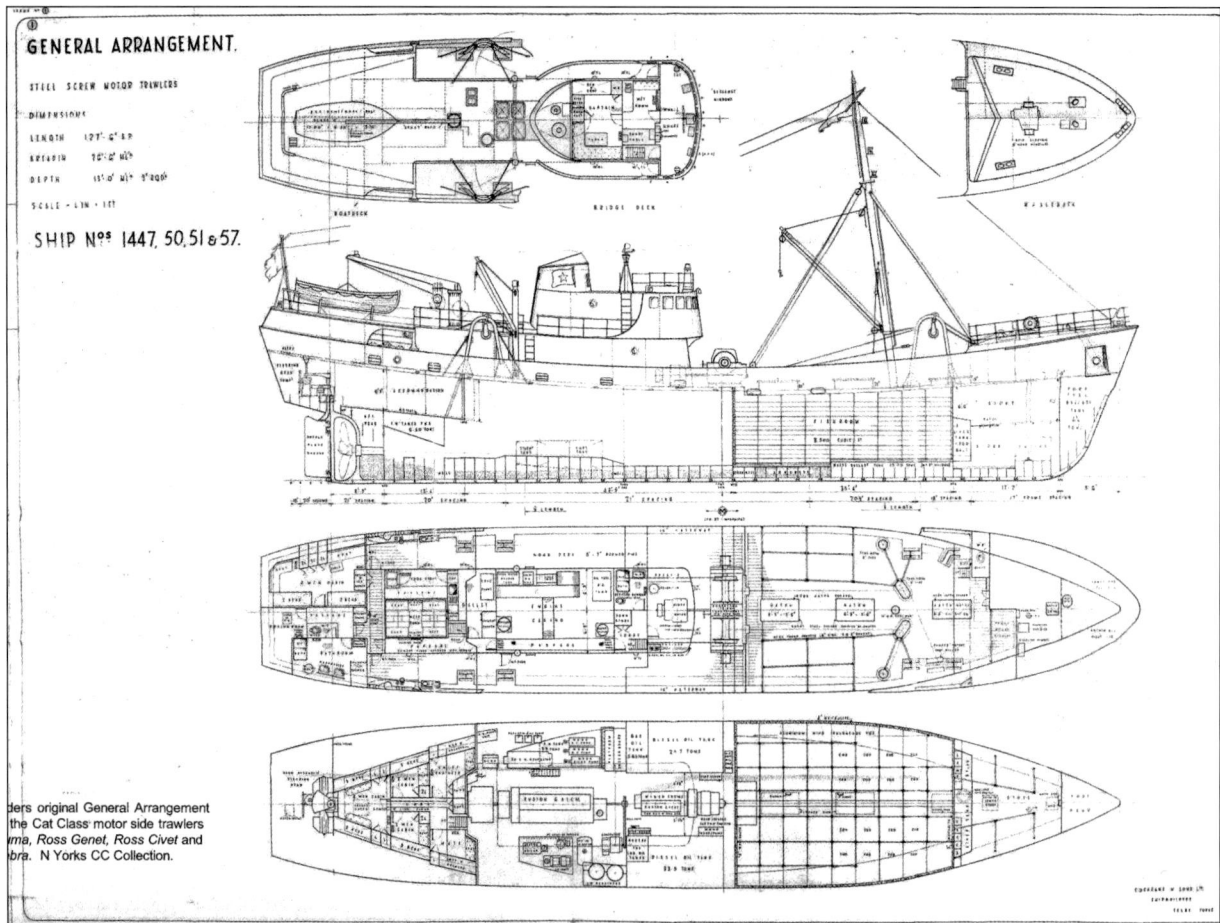

A general arrangement drawing for yard numbers 1447, 1450, 1451 and 1457.

(Cochrane archive)

ROSS CIVET	1451	352	127.5	Ruston & Hornsby	Ross Trawlers Ltd,
301842	12.05.1960	116	26.5	1025bhp 6-cyl	Grimsby
GY652 Motor trawler	02.11.1960		13.0	11.0 knots	

Cost to build £153,555 (WFA Grant £37,500).
27.10.1960: Registered at Grimsby (GY652). 18.10.1961: Sold to Dennis Roberts Fishing Co Ltd, Grimsby. 31.03.1966: Sold to Ross Trawlers Ltd, Grimsby. 01.07.1969: Became part of the British United Trawlers (B.U.T.) fleet. B.U.T. livery adopted. Book value £73,065. 30.09.1969: Sold to Goweroak Ltd, Grimsby. 29.09.1982: Sold to British United Trawlers (Grimsby) Ltd, Grimsby. 07.02.1985: Sold to Cam Shipping Co Ltd, Hull. Converted to offshore platform standby safety vessel by Globe Engineering Ltd, Hull. 30.04.1985: Registered at Grimsby as **CAM CIVET** (IMO 5300338). 06.1992: Laid up following new regulations for offshore platform standby safety vessels. 11.1993: Sold to Bideawhile One Hundred and Twenty Nine Ltd, Ipswich. 1999: Sold to Empresa de Pesca Ribau, São Tomé, São Tomé & Principe, Gulf of Guinea, (based in Ilhavo, Portugal). 2000: Re-engined with 8-cylinder 1014bhp oil engine (made 1985) by Klöckner-Humbolt-Deutz AG, Cologne. Remeasured to 398gt, 119n. 11.1993: Grimsby registry closed. Registered at São Tomé. 2015: Believed to be still in service.

Not built	1452 Order cancelled				

DAYSPRING	1453	453	137.0	Mirrlees	Robins Trawlers Ltd,
301628	19.09.1959	150	28.0	1000bhp 8-cyl	Hull
H183 Motor trawler	01.03.1960		14.3	11.5 knots	

Cost to build £182,000 (WFA Grant £37,000). Mortgage provided by the WFA, London.
29.02.1960: Registered at Hull (H183). 03.03.1960: Sailed on first trip to Faroe grounds. 23.06.1961: Last landing at Hull; laid up and offered for sale. 08.1961: Company bought by Basil Arthur Parkes, Fleetwood. 15.08.1961: Sold to Parbel Smith Ltd, Aberdeen (Boston Deep Sea Fisheries Group, Fleetwood). 28.11.196: Sailed Fleetwood for fishing grounds (Sk Ronald Slapp) in fine clear weather. At 9.00am off Isle of Man in collision with London-registered tanker AUSPICITY (402grt/1944) on passage from Greenock towards Liverpool with part cargo of vegetable oil; both vessels damaged. AUSPICITY listing on port side rails and shell plates midships set in and holed. DAYSPRING returned to Fleetwood leaking in forepeak. 29.11.1961: After landing ice, placed on slip for survey and repair. Stern twisted, several shell plates set in and holed on starboard bow. Repairs effected and returned to service. 03.04.1962: Registered at Hull as **ADMIRAL NELSON** (H183). Fishing out of Fleetwood. 25.04.1962: Transferred to fish out of Grimsby. 09.03.1963: Registered at Hull as **PRINCESS ROYAL** (H183). 29.08.1968: Sold to Irvin & Johnson Ltd, Cape Town, South Africa. 18.10.1968: Hull registry closed. 10.1968: Registered at Cape Town. 1983: Vessel stripped of reusable parts and non-ferrous metals. 09.05.1983: Hulk scuttled off Robben Island, Cape Town. 05.1983: Cape Town registry closed.

Not built	1454, 1455 Orders cancelled				

LAVINDA	1456	285	115.0	Mirrlees	J Marr & Son Ltd,
301885	15.01.1960	93	25.0	750bhp 6-cyl	Fleetwood
FD159 Motor trawler	23.06.1960		12.5	12.0 knots	

Cost to build £147,200. Built with the aid of a WFA Grant.
20.06.1960: Registered at Fleetwood (FD159). 14.04.1964: Sold to Richard Irvin & Sons Ltd Aberdeen, for the sum of £124,000. 24.04.1964: Registered at Fleetwood as **BEN ARTHUR** (FD159). 02.05.1964: Fleetwood registry closed. 05.1964: Registered at Aberdeen (A742). 1966: Lengthened to 119.3ft and remeasured to 307grt, 113net. 23.04.1971: Sold to Gunnar Hafsteinsson, Reykjavik, Iceland. 04.1971: Aberdeen registry closed. 04.1971: Registered at Reykjavik as **FREYJA** (RE38). 16.07.1975: Sold to Ársæll s/f, Hafnarfjörður. Reykjavik registry closed. Registered at Hafnarfjörður as **ÁRSÆLL SIGURÐSSON II** (HF12). 1980: Re-registered at Hafnarfjörður as (HF120) as owner purchased the stern trawler DAGNY (385grt/1966) which they wished to rename ÁRSÆLL SIGURÐSSON (H12). 04.12.1980: Declared unseaworthy, taken off the register and laid up in the outer harbour at Reykjavik. 10.1984: Sold to Albert Draper & Son Ltd, Hull. 27.10.1984: Arrived Hull under tow and broken up at the Victoria Dock slipway.

ROSS ZEBRA	1457	352	127.5	Ruston & Hornsby	Ross Trawlers Ltd,
301845	11.06.1960	116	26.5	1025bhp 6-cyl	Grimsby
GY653 Motor trawler	29.11.1960		13.0	11.0 knots	

Cost to build £153,555 (WFA Grant £37,500).
28.11.1960: Registered at Grimsby (GY653). 30.09.1961: Sold to Andanes Steam Fishing Co Ltd, Grimsby. 31.03.1966: Sold to Ross Trawlers Ltd, Grimsby. 01.07.1969: Became part of the British United Trawlers (B.U.T.) fleet. B.U.T. livery adopted. Book value £72,438. 30.09.1969: Sold to Goweroak Ltd, Grimsby. 29.09.1982: Sold to British United Trawlers (Grimsby) Ltd, Grimsby. 06.02.1985: Sold to Cam Shipping Co Ltd, Grimsby. Converted to offshore platform standby safety vessel by Globe Engineering Ltd, Hull. 29.04.1985: Registered at Grimsby as **CAM ZEBRA**. 06.1992: Laid up following new regulations for offshore platform standby safety vessels. 11.1993: Sold to Bideawhile One Hundred and Twenty Nine Ltd, Ipswich. 2000: Sold to undisclosed owners. 2000: Grimsby registry closed. Registered at Lomé, Togo, as **MILAGRE**. 02.2004: Sold to Ivoni Fishing Co Ltd, Ghana. Togo registry closed. Registered at Takoradi as **IVONI I** (IMO 5300716). 2015: Believed to still be in service.

CLIONE	1458	495	135.0	Ruston & Hornsby	Ministry of Agriculture,
302394 IMO 5076119	24.08.1960	129	28.0	1060bhp 6-cyl	Fisheries & Food,
LT421 Research ship	15.03.1961		14.25	11.5 knots	Lowestoft

03.1961: Registered at Lowestoft. 30.01.1987: Sold to Putford Enterprises Ltd, Lowestoft. 13.04.1988: Registered at Lowestoft as **PUTFORD PETREL**. 1992: Converted to offshore platform standby safety vessel. 02.03.2000: Sold to William Barbour, Lowestoft. Registered at Lowestoft as **LYNN G**. 04.2000: Sold to Dutch Film Unit BV, Lete, Netherlands, & Mr William Barbour, Chelmsford. Converted to a private vessel and classed as a yacht. Remeasured to 482gt. 2015: Still in service.

Ross Civet (1451) when new.

(Cochrane archive)

Lavinda (1456) at full speed.

(Cochrane archive)

Clione (1458) at anchor.

(Cochrane archive)

ROSS EAGLE	1459	288	107.5	Ruston & Hornsby	Ross Trawlers Ltd,
301850	22.09.1960	92	24.5	550bhp 5-cyl	Grimsby
GY656 Motor trawler	11.04.1961		12.5	12.0 knots	

Cost to build £122,811 (WFA Grant £30,703).
11.04.1961: Registered at Grimsby (GY656).   20.09.1961: Sold to T C & F Moss Ltd, Grimsby.   31.03.1966: Sold to Ross Trawlers Ltd, Grimsby.   30.09.1969: Sold to Goweroak Ltd, Grimsby.   01.07.1969: Became part of the British United Trawlers (B.U.T.) fleet. B.U.T. livery adopted. Book value £61,474.   05.1975: Converted to offshore platform standby safety vessel.   26.10.1979: Remeasured to 272grt, 86net.
02.05.1982: Sold to George Craig & Sons Ltd, Aberdeen.   02.06.1882: Sold to The Colne Shipping Co Ltd, Luton.   14.07.1982: Grimsby registry closed.   21.07.1982: Registered at Lowestoft.   02.01.1983: Registered at Lowestoft as **INAGUA**.   1992: Sold to C Knight/Caravel Maritime Ltd, Rainham, Kent, in an en bloc deal with NOTTS FOREST (409grt/1960); COLNE PHANTOM (405grt/1965); HUDDERSFIELD TOWN (408grt/1962) and EXUMA (165grt/1961).   13.01.1992: Sailed Lowestoft under own power for River Medway.   14.01.1992: Delivered for breaking up by Caravel Maritime Ltd at Bloors Wharf, Rainham. Registry closed.

ROSS HAWK	1460	288	107.5	Ruston & Hornsby	Ross Trawlers Ltd,
301854	24.10.1960	92	24.5	550bhp 5-cyl	Grimsby
GY657 Motor trawler	16.05.1961		12.5	12.0 knots	

Cost to build £123,761 (WFA Grant £30,940).
15.05.1961: Registered at Grimsby (GY657).   30.09.1961: Sold to George F Sleight & Sons Ltd, Grimsby.   30.09.1967: Sold to Ross Trawlers Ltd, Grimsby.   01.07.1969: Became part of the British United Trawlers (B.U.T.) fleet. B.U.T. livery adopted. Book value £61,554.
30.09.1969: Sold to Goweroak Ltd, Grimsby.   03.11.1971: When some 4 miles NNW of Holborn Head disabled with engine problems. Located by Thurso lifeboat THE THREE SISTERS, connected and delivered Scrabster.   12.1975: Laid up at Granton.   06.1976: Converted to offshore platform standby safety vessel.   21.05.1982: Sold to George Craig & Sons, Aberdeen.   02.06.1982: Sold to The Colne Shipping Ltd, Luton.   14.07.1982: Grimsby registry closed.   21.07.1982: Registered at Lowestoft.   01.08.1983: Registered at Lowestoft as **PAGONA**.
03.1991: Sold to Chequers Iron & Steel Ltd, Dagenham, for breaking up.   12.03.1991: Sailed Lowestoft in tow of DOMINICA (see yard no.1465) also for breaking up.   13.03.1991: Delivered River Medway for breaking up at Bloors Wharf, Rainham, by Tollgate Metals Ltd.
02.05.1991: Registry closed.

ROSS KESTREL	1461	288	107.6	Ruston & Hornsby	Ross Trawlers Ltd,
303271	21.11.1960	92	24.0	550bhp 5-cyl	Grimsby
GY658 Motor trawler	13.06.1961		12.6	12.0 knots	

Cost to build £124,830 (WFA Grant £31,207).
12.06.1961: Registered at Grimsby (GY658).   30.09.1961: Sold to George F Sleight & Sons Ltd, Grimsby.   01.07.1969: Became part of the British United Trawlers (B.U.T.) fleet. B.U.T. livery adopted. Book value £62,251.   30.09.1967: Sold to Ross Trawlers Ltd, Grimsby.
30.09.1969: Sold to Goweroak Ltd, Grimsby.   1970: Fishing out of Fleetwood.   1972: Fishing out of Granton.   27.05.1973: Fishing 12 miles east off Fair Isle (Sk Joey Jamieson) brought a live torpedo up in the trawl. Sk Jamieson thought it safer to secure the torpedo onboard and leave disposal to naval authorities on return to Granton. Torpedo subsequently exploded killing Chief Engineer Robert Clements and 2nd Engineer James Maruszczak. Ship damaged by explosion, crew abandoned and picked up from the liferafts by the Granton trawler ARCTIC EXPLORER (GW7) (273grt/1957) and transferred off Sumburgh Head to the Lerwick lifeboat.   02.1975: Converted to offshore platform standby safety vessel.   1979: Laid up for disposal.   1979: Sold to J Gibson Johnstone Ltd, Hull.   1979: Sold to H Kitson Vickers & Sons (Engineering) Ltd, Sheffield, then re-sold to S Dalton Skip Hire Ltd, Edinburgh.   21.11.1979: Arrived Bo'ness for breaking up.
07.10.1981: Grimsby registry closed.

ROSS RENOWN	1462	790	185.0	Ruston & Hornsby	Hudson Bros (Trawlers) Ltd,
303288	28.08.1961	283	33.0	2160bhp 6-cyl	Grimsby
GY666 Motor trawler	23.03.1962		17.25	CPP 13.0 knots	

Cost to build £307,905.
21.03.1962: Registered at Grimsby (GY666).   29.09.1962: Sold to George F Sleight & Sons Ltd, Grimsby.   14.03.1966: Sold to Queen Steam Fishing Co Ltd, Grimsby.   30.09.1967: Sold to Ross Trawlers Ltd, Grimsby.   01.04.1968: In a NW gale and heavy seas running, responded with trawler WILLIAM WILBERFORCE (H200) (698grt/1959) to distress call from Grimsby trawler ROSS PUMA (see yard no.1447) hard aground on the Little Rackwick shoals beneath the cliffs two miles NNW of Tor Ness, Hoy. Orkney Islands, but could not approach casualty. Stood by while Longhope lifeboat picked up crew from liferafts.   01.07.1969: Became part of the British United Trawlers (B.U.T.) fleet. B.U.T. livery adopted. Book value £186,833.   01.12.1976: Last Grimsby trawler to leave the Icelandic fishing grounds following the imposition of the 200 mile fishing limit around Iceland.   08.1979: Sold to Hughes Bolckow Ltd, Blyth.   08.1979: Arrived Blyth for breaking up.   29.09.1979: Grimsby registry closed.

PRISCILLIAN	1463	302	120.0	Drypool/Brons	Dominion Steam Fishing
303280	18.03.1961	95	25.6	790bhp 12-cyl	Co Ltd,
GY672 Motor trawler	09.10.1961		10.9	CPP 12.7 knots	Grimsby

Built with the aid of WFA Grant.
02.10.1961: Registered at Grimsby (GY672). Insured for £155,000.   10.1976: Owner ceased trading.   13.12.1976: Sold to Small & Co (Lowestoft) Ltd, Lowestoft.   10.01.1977: Grimsby registry closed.   13.01.1977: Registered at Lowestoft 09.03.1977: Registered at Lowestoft as **SUFFOLK CRAFTSMAN** (LT372).   1980: Sold to John Xiros, Athens, Greece.   09.06.1980: Lowestoft registry closed. Registered at Athens as **ION**.   07.1981: Sold to Th Bakopoulos S.A., Athens, and broken up at Athens. Registry closed.

TIBERIAN	1464	302	120.0	Drypool/Brons	Sir Thomas Robinson & Son
303281	02.05.1961	120	25.6	790bhp 12-cyl	(Grimsby) Ltd,
GY673 Motor trawler	14.11.1961		10.6	CPP 12.7 knots	Grimsby

Built with the aid of WFA Grant. Insured for £155,000.
26.10.1961: Registered at Grimsby (GY673).   10.1976: Owner ceased trading.   19.10.1976: Sold to Dagon Fishing Co Ltd, Luton.
29.10.1976: Arrived Lowestoft.   30.11.1976: Grimsby registry closed.   03.12.1976: Registered at Lowestoft (LT349).   20.01.1977: Surveyed after conversion to offshore platform standby safety vessel.   22.08.1978: Sold to Small & Co (Lowestoft) Ltd, Lowestoft.
28.11.1978: Registered at Lowestoft as **SUFFOLK MAID** (LT349). Converted back to fishing.   03.05.1980: Sold to Jack Elmore, Mirfield, Yorkshire.   12.06.1980: Lowestoft registry closed - "no longer fishing".   1981: Sold to J Elinour, Christiansted, US Virgin Islands.   1981: At Saint Croix, US Virgin Islands, to be converted to a cargo vessel but due to difficulties, laid up in Frederiksted harbour.   07.11.1984: Parted her moorings during hurricane "Klaus" and drove on to Frederiksted Pier.   12.1985: Refloated after several attempts. Superstructure removed, taken out and scuttled 1 mile off in Butler Bay, St. Croix, to form artificial reef/dive site. Vessel upright in clear water at 50ft/15,24m.

ROSS CORMORANT	1465	288	107.5	Ruston & Hornsby	Ross Trawlers Ltd,
303275	03.01.1961	92	24.5	550bhp 5-cyl	Grimsby
GY665 Motor trawler	25.07.1961		12.5	12.5 knots	

Cost to build £125,836 (WFA Grant £31,459).
20.07.1961: Registered at Grimsby (GY665). 30.09.1961: Sold to George F Sleight & Sons Ltd, Grimsby. 01.07.1969: Became part of the British United Trawlers (B.U.T.) fleet. B.U.T. livery adopted. Book value £62,913. 30.09.1967: Sold to Ross Trawlers Ltd, Grimsby. 30.09.1969: Sold to Goweroak Ltd, Grimsby. 03.03.1970: Transferred to fish out of Fleetwood. 1973: Fishing out of Aberdeen. 1979: Converted to offshore platform standby safety vessel. 21.05.1982: Sold to George Craig & Sons Ltd, Aberdeen. 02.06.1982: Sold to Colne Shipping Ltd, Luton. 14.07.1982: Grimsby registry closed. 02.09.1982: Registered at Lowestoft as **DOMINICA**. 23.02.1984: Surveyed for offshore platform standby safety role. 03.1991: Sold to Chequers Iron & Steel Ltd, Dagenham, for breaking up. 12.03.1991: Sailed Lowestoft towing PAGONA (see yard no.1460) also for breaking up. 13.03.1991: Delivered River Medway for breaking up at Bloors Wharf, Rainham, by Tollgate Metals Ltd. 05.12.1991: Registry closed - "Vessel broken up".

ROSS FALCON	1466	288	107.5	Ruston & Hornsby	Ross Trawlers Ltd,
303277	17.02.1961	92	24.5	550bhp 5-cyl	Grimsby
GY667 Motor trawler	05.09.1961		12.5	12.5 knots	

Cost to build £126,397 (WFA Grant £31,599).
28.08.1961: Registered at Grimsby (GY667). 30.09.1961: Sold to George F Sleight & Sons Ltd, Grimsby. 30.09.1967: Sold to Ross Trawlers Ltd, Grimsby. 01.07.1969: Became part of the British United Trawlers (B.U.T.) fleet. B.U.T. livery adopted. Book value £65,100. 30.09.1969: Sold to Goweroak Ltd, Grimsby. 06.1975: Sold to NV Rederij Noordvisserij, Ostend, Belgium. 11.06.1975: Grimsby registry closed. Converted to a beam trawler and re-engined with a 6-cylinder 1000bhp oil engine by Anglo-Belgian Co B.V., Gent. 12.12.1975: Registered at Ostend as **FALCON** (O.313). 27.12.1984: Ostend registry closed. 22.03.1985: Arrived Bruges for breaking up by J Bakker & Zonen. Registry closed.

ROSS KITTIWAKE	1467	288	107.5	Ruston & Hornsby	Ross Trawlers Ltd,
303284	01.06.1961	92	24.5	550bhp 5-cyl	Grimsby
GY678 Motor trawler	07.12.1961		12.5	12.5 knots	

Cost to build £127,256 (WFA Grant £31,814).
06.12.1961: Registered at Grimsby (GY678). 29.09.1962: Sold to George F Sleight & Sons Ltd, Grimsby. 30.09.1967: Sold to Ross Trawlers Ltd, Grimsby. 01.07.1969: Became part of the British United Trawlers (B.U.T.) fleet. B.U.T. livery adopted. Book value £64,574. 30.09.1969: Sold to Goweroak Ltd, Grimsby. 1979: Converted to offshore platform standby safety vessel. 21.05.1982: Sold to George Craig & Sons Ltd, Aberdeen. 02.06.1982: Sold to The Colne Shipping Co Ltd, Luton. 12.07.1982: Grimsby registry closed. 01.09.1982: Registered at Lowestoft as **DESIRADE**. 08.1991: Sold to Spearing International Ltd, Dartford, in an en bloc deal with BERMUDA (see yard no.1478); SCAMPTON (213grt/1962) and WADDINGTON (212grt/1962). 06.09.1991: Delivered Medway for breaking up at Bloors Wharf, Rainham, by Caravel Maritime Ltd. 12.01.1993: Registry closed - "Vessel broken up".

ROSS HERON	1468	288	107.5	Ruston & Hornsby	Ross Trawlers Ltd,
303290	25.10.1961	92	24.5	695bhp 6-cyl	Grimsby
GY693 Motor trawler	17.04.1962		12.5	12.5 knots	

Cost to build £131,000 (WFA Grant £31,000).
02.05.1962: Registered at Grimsby (GY693). 01.07.1969: Became part of the British United Trawlers (B.U.T.) fleet. B.U.T. livery adopted. Book value £67,149. 30.09.1969: Sold to Goweroak Ltd, Grimsby. 03.03.1970: Transferred to fish out of Fleetwood. 11.11.1971: Fishing out of Granton. 17.08.1981: Sold to Colne Fishing Co Ltd, Luton. 1981: Converted to offshore platform standby safety vessel. 14.09.1981: Grimsby registry closed. 18.09.1981: Registered at Lowestoft. 10.02.1982: Registered at Lowestoft as **MUSTIQUE** (exchanged names with existing MUSTIQUE (see yard no.1397 laid up with engine damage). 07.1991: Sold to Spearing International Ltd, Dartford, in an en bloc deal with COLNE HUNTER (442grt/1963). 29.07.1991: Sailed Lowestoft for River Medway, MUSTIQUE towing COLNE HUNTER. 10.08.1991: Delivered Medway for breaking up at Bloors Wharf, Rainham, by Masterman Iron & Steel Shipbreakers Ltd. Registry closed.

ROSS CURLEW	1469	288	107.5	Ruston & Hornsby	Ross Trawlers Ltd,
303293	24.11.1961	92	24.5	695bhp 6-cyl	Grimsby
GY692 Motor trawler	23.05.1962		12.5	12.5 knots	

Cost to build £131,000 (WFA Grant £31,000).
21.05.1962: Registered at Grimsby (GY692). 01.07.1969: Became part of the British United Trawlers (B.U.T.) fleet. B.U.T. livery adopted. 30.09.1969: Sold to Goweroak Ltd, Grimsby. 21.07.1970: Transferred to fish out of Fleetwood. 11.1971: Transferred to fish out of Granton. 1978: Fishing out of Aberdeen. 29.06.1981: Sold to Colne Fishing Co Ltd, Lowestoft. 1981: Converted to offshore platform standby safety vessel. 07.09.1981: Grimsby registry closed. 09.09.1981: Registered at Lowestoft. 17.09.1981: Registered at Lowestoft as **ANEGADA**. 12.1991: Sold to C Knight/Caravel Maritime Ltd, Rainham, in an en bloc deal with BARNSLEY (408grt/1960); COLNE KESTREL (405grt/1966) and CRYSTAL PALACE (410grt/1962). 20.12.1991: Arrived River Medway for breaking up at Bloors Wharf, Rainham, by Caravel Maritime Ltd. Registry closed.

STELLA ORION	1470	778	186.0	Mirrlees	Ross Trawlers Ltd,
301699	08.03.1962	260	33.5	1800bhp 6-cyl	Hull
H235 Motor trawler	24.07.1962		17.25	15.0 knots	

Cost to build £330,274.
23.07.1962: Registered at Hull (H235). 28.07.1962: Sailed on first trip – White Sea. 29.03.1963: Sold to Hudson Brothers Trawlers Ltd, Hull. 03.01.1966: Registered at Hull as **ROSS ORION** (H235). Ross Group funnel colours adopted. 1968: 3rd place in The Silver Cod Trophy with several skippers, 39,806 kits - 337 days at sea - £166,215 gross. 01.07.1969: Became part of the British United Trawlers (B.U.T.) fleet. B.U.T. livery adopted. Book value £188,871. 1975: Runner up in The Hull Challenge Shield Competition with skipper Alfred Osler, 33,066 kits. 18.01.1978: Landed at Hull. Laid up for disposal due to fishing limits and increased fuel price. 13.08.1979: Sold to Hughes Bolckow Ltd, Blyth, for breaking up. 17.12.1979: Hull registry closed - "Vessel broken up".

Ross Hawk (1460)

(Cochrane archive)

Ross Curlew (1469) alongside at Grimsby for the first time.

(Jonathan Grobler collection)

Priscillian (1463)

(Cochrane archive)

Not built	1471 to 1477 Orders cancelled				
ROSS MALLARD 304775 GY699 Motor trawler	1478 06.04.1962 11.09.1962	288 92	107.5 24.5 12.5	Ruston & Hornsby 695bhp 6-cyl 12.5 knots	Ross Trawlers Ltd, Grimsby

Cost to build £136,000 (No WFA Grant available).
06.09.1962: Registered at Grimsby (GY699). 01.10.1962: Sold to Alsey Steam Fishing Co Ltd, Grimsby. 04.1964: Chartered by University of Hull for sand wave recording and mapping experiments in the vicinity of Flamborough Head. 01.10.1965: Sold to Ross Trawlers Ltd, Grimsby. 01.07.1969: Became part of the British United Trawlers (B.U.T.) fleet. B.U.T. livery adopted. Book value £92,247.
01.09.1970: Sold to George F Sleight & Sons Ltd, Grimsby. 20.08.1971: Sold to British United Trawlers (Granton) Ltd, Granton, Edinburgh.
11.09.1971: Transferred to fish out of Granton. 20.05.1977: Sold to Strathcoe Fishing Co Ltd, Aberdeen. Transferred to fish out of Aberdeen. 30.09.1978: Sold to British United Trawlers (Aberdeen) Ltd, Aberdeen. 06.06.1980: Sold to Drifter Trawlers Ltd, Lowestoft.
1980: Converted to offshore platform standby safety vessel. 23.07.1980: Grimsby registry closed. 29.07.1980: Registered at Lowestoft.
17.11.1980: Registered at Lowestoft as **BERMUDA**. 06.12.1980: Surveyed following conversion to offshore platform strandby safety vessel.
31.12.1987: Sold to The Colne Shipping Co Ltd, Lowestoft. 08.1991: Sold to C Knight/CaravelMaritime Ltd, Rainham, in an en bloc deal with DESIRADE (see yard no.1467); SCAMPTON (213grt/1962) and WADDINGTON (212grt/1962). 22.08.1991: Sailed Lowestoft for River Medway. 23.08.1991: Arrived River Medway for breaking at Bloors Wharf, Rainham, by Caravel Maritime Ltd. Offered for sale.
08.1991: Chartered by Capt D Potter, Hull. 10.1993: On completion of charter, commenced breaking up at Bloors Wharf. Registry closed.

ROSS TERN 304778 GY700 Motor trawler	1479 05.05.1962 10.10.1962	288 92	107.5 24.5 12.5	Ruston & Hornsby 785bhp 6-cyl 12.5 knots	Ross Trawlers Ltd, Grimsby

Cost to build £138,000 (No WFA Grant available).
10.10.1962: Registered at Grimsby (GY700). 24.10.1962: Sold to Kopanes Steam Fishing Co Ltd, Grimsby. 06.1965: Converted to semi unmanned engine room. 31.03.1966: Sold to Queen Steam Fishing Co Ltd, Grimsby. 01.07.1969: Became part of the British United Trawlers (B.U.T.) fleet. B.U.T. livery adopted. Book value £93,333. 16.06.1970: Sold to Northern Trawlers Ltd, Grimsby. 07.06.1971: Sold to George F Sleight & Sons Ltd, Grimsby. 16.06.1972: Sold to British United Trawlers (Aberdeen) Ltd, Aberdeen. 09.02.1973: Sailed Aberdeen at 11.00am for West of Scotland grounds (Sk William M M Gardner). In a moderate NW wind, heavy snow showers and confused sea, at 11.30pm stranded on Tarf Tail (little Windy Skerry), SE end of Swona, Pentland Firth. Crew picked up by Longhope lifeboat DAVID AND ELIZABETH KING (Cox John Leslie), six from liferaft and five from the ship. 12.02.1973: Declared a total loss. 06.03.1973: Grimsby registry closed. 1973: Cox Leslie awarded RNLI Bronze medal for the rescue of the eleven crewmen.

CAPE CANAVERAL 303823 H267 Motor trawler	1480 30.10.1962 02.04.1963	805 275	186.0 33.5 17.25	Mirrlees National 1800bhp 6-cyl 13.2 knots	Ross Trawlers Ltd, Grimsby

Cost to build £341,506. Government Grant £50,000.
02.04.1963: Registered at Hull (H267). 04.04.1963: Sailed on first trip. 01.10.1963: Sold to Hudson Brothers Trawlers Ltd, Hull.
29.11.1965: Registered at Hull as **ROSS CANAVERAL** (H267). Ross Group funnel colours adopted. 1965: The Silver Cod Trophy placed third with skipper George J D Thompson Whur - 34,172 kits, 347 days at sea, £149,643 gross. 01.07.1969: Became part of the British United Trawlers (B.U.T.) fleet. B.U.T. livery adopted. Book value £197,910. 1970: Runner-up in The Hull Challenge Shield Competition with Sk Ken Neilson - 34,009 kits, £207,151gross. 15.11.1977: Outwards for fishing grounds stranded off Godoy Island, refloated and towed to Aalesund for survey and temporary repairs. Towed back to Hull. Full survey indicated vessel beyond economical repair. Laid up for disposal. 23.06.1978: Sold to Tees Marine Services Ltd, Middlesbrough. 14.07.1978: Arrived Middlesborough in tow for breaking up.
22.11.1978: Fishing registration closed. 27.11.1978: Hull registry closed - "Vessel broken up".

Not built	1481 Order cancelled				
AVENGER 304381 IMO 5401297 Motor tug	1482 20.06.1962 16.11.1962	283	110.0 30.0 14.5	British Polar 1800bhp 9-cyl 12.0 knots	Elliott Steam Tug Co Ltd, London

14.11.1962: Registered at London. 03.1965: Sold to William Watkins Ltd, London. 20.09.1968: Owners restyled London Tugs Ltd.
1975: Fitted with Kort nozzle & controlled pitch propeller by Tyne Dock & Engineering Co Ltd, South Shields. 01.01.1975: Company and assets sold to The Alexandra Towing Co Ltd, Liverpool. 28.04.1976: Owners restyled The Alexandra Towing Co (London) Ltd, London.
1985: Sold to Purvis Marine, Sault Ste Marie, Ontario, Canada. London registry closed. Registered at Sault Ste Marie as **AVENGER IV**.
31.10.1985: Sailed Gravesend bound for the Great Lakes. 2015: Still in service.

HIBERNIA 304415 IMO 5405085 Motor tug	1483 17.08.1962 16.01.1963	293	110.0 30.0 14.5	British Polar 1800bhp 9-cyl 12.0 knots	William Watkins Ltd, London

01.1963: Registered at London. 20.09.1968: Owners restyled London Tugs Ltd. 17.11.1973: From Dover took out Kent Fire Brigade firefighters and equipment to German motor vessel CAP SAN ANTONIO (7636grt/1972) on fire off Shakespeare Cliffs. Cargo included dangerous chemicals. 01.01.1975: Company and assets sold to The Alexandra Towing Co Ltd, Liverpool. 28.04.1976: Owners restyled The Alexandra Towing Co (London) Ltd, London. 1975: Fitted with Kort nozzle & controlled pitch propeller. 1984: Classed as a fire-fighting tug. 09.1987: Sold to Achilleus II Shipping Co, Thessaloniki, Greece. London registry closed. Registered at Thessalonika as **ATROTOS**.
06.1991: Sold to Karapiperis X Tug & Salvage Shipping Co, Piræus. Thessalonika registry closed. Registered at Piræus as **KARAPIPERIS X**. 07.06.2006: Sold to Katakolon Tugboat Services Shipping Co, Katakolon, Greece. 01.2007: Registered at Piræus as **ALFIOS**. 10.11.2007: In heavy seas and gale force winds responded to distress call from Turkish motor vessel AKDAG (393gt/1978) disabled with engine problems some 12 miles south of Katakolon. Connected and delivered Katacolon - 293 immigrants discovered in hold.
2011: Sold to Ergasies Rimoulkiseos Katakolou Naftiki Eteria, Piræus. 2015: Still in service.

An impressive stern view of **Cape Canaveral** (1480) on trials.

(Cochrane archive)

Another splendid trials image of the same trawler.

(Cochrane archive)

The wheelhouse of **Cape Canaveral**.

(Cochrane archive)

Ross Tern (1479)

(Cochrane archive)

A trials view of the firefighting tug **Avenger** *(1482) in the colours of the Elliott Steam Tug Co Ltd.*

(Cochrane archive)

Sister tug **Hibernia** *(1483) also on trials but carrying the funnel colours of William Watkins Ltd.*

(Cochrane archive)

STELLA SIRIUS 303836 H277 Motor trawler	1484 12.03.1963 16.07.1963	677 225	175.8 32.5 16.75	Holmes Werkspoor 1622bhp 8-cyl 14.0 knots	Hudson Bros Trawlers Ltd, Hull

Cost to build £312,685.
18.07.1963: Registered at Hull (H277).   20.07.1963: Sailed on first trip.   29.11.1965: Registered at Hull as **ROSS SIRIUS** (H277).   Ross Group funnel colours adopted.   1968: Runner up in The Hull Challenge Shield Competition with several skippers - 39,569 kits, £159,376 gross.   01.07.1969: Became part of the British United Trawlers (B.U.T.) fleet.   Adopted B.U.T. livery.   Book value £199,368.
1977: Runner-up in The Hull Challenge Shield Competition with Sk Denis Whiting - 33,066 kits, £544,815 gross, £284,292 net.
10.07.1978: Landed and laid up for disposal, due to fishing limits and increased fuel price.   26.10.1979: Fishing registration closed.
.04.1981: Sold to Albert Draper & Son Ltd, Hull.   07.04.1981: Delivered Victoria Dock slipway for breaking up.   21.05.1981: Hull registry closed.

STELLA ALTAIR 303839 H279 Motor trawler	1485 24.03.1963 24.09.1963	677 225	175.8 32.5 16.75	Holmes Werkspoor 1622bhp 8-cyl 14.0 knots	Hudson Bros Trawlers Ltd, Hull

Cost to build £312,543.   Last side fishing trawler built for the Hull fishing fleet.
18.09.1963: Registered at Hull (H279).   28.09.1963: Sailed on first trip.   29.11.1965: Registered at Hull as **ROSS ALTAIR** (H279).   Ross Group funnel colours adopted.   01.07.1969: Became part of the British United Trawlers (B.U.T.) fleet.   Adopted B.U.T. livery.   Book value £202,797.   1973: Winner of The Hull Challenge Shield with Sks Neville Beavers and Benjamin Ashcroft - 25,352 kits, £343,679 gross.
13.07.1978: Landed and laid up for disposal, due to fishing limits and increased fuel price.   26.10.1979: Fishing registration closed.
04.1981: Sold to Albert Draper & Son Ltd, Hull.   08.04.1981: Delivered Victoria Dock slipway for breaking up.   21.05.1981: Hull registry closed.

HEADMAN 303825 IMO 5411486 Twin screw motor tug	1486 13.12.1962 28.03.1963	193	95.0 28.0	Lister Blackstone 1320bhp 2 x 8-cyl 11.0 knots	United Towing Co Ltd, Hull

29.03.1963: Registered at Hull.   01.04.1978: Sold to Humber Tugs Ltd, Grimsby.   08.1981: Sold to Alexander G Tsavliris & Sons, Piræus, Greece.   01.09.1981: Hull registry closed.   Registered at Piræus as **HERMES**.   Later company restyled as Tsavliris & Sons Maritime Ltd, Piræus.   18.11.2011: On salvage station at Patras.   Attended the Panamanian flag motor vessel FGM COMMANDER (6819gt/1985) stranded on shallow rocky area in the Aegean Sea, 2.5 miles south of Dokos Island, while on passage Elefsina, Greece, towards Béjaïa, Algeria, with steel bars.   23.11.2011: After lightering operations, connected and with salvage tug ALEXANDER 3 (545gt/1989) successfully refloated FGM COMMANDER which was delivered to anchorage for underwater inspection.   2015: Still in service.

WORKMAN 303828 IMO 5414397 Twin screw motor tug	1487 11.01.1963 01.05.1963	193	95.0 28.0	Lister Blackstone 1320bhp 2 x 8-cyl 11.0 knots	United Towing Co Ltd, Hull

23.04.1963: Registered at Hull.   21.01.1970: Owners restyled as United Towing Ltd.   17.03.1978: Company restyled as North British Maritime Group Ltd, Hull.   01.04.1978: Sold to Humber Tugs Ltd, Grimsby.   16.01.1979: Sold to Wenhove Ltd, Aberdeen.   19.04.1979: Hull registry closed.   Registered at Aberdeen as **DUNCAN**.   1979: Sold to Ocean Bed Salvage Co Ltd, Douglas, Isle of Man.   02.1980: For sale.
1980: Sold to Frank Pearce (Tugs) Ltd, Poole.   Aberdeen registry closed.   Registered at Poole as **PULLWELL DELTA**.   29.01.1981: Sailed Glasgow for River Thames towing Clyde passenger steamer QUEEN MARY (1014gt/1933).   02.02.1981: Arrived King George V dock, London.   27.02.1983: Arrived Gateshead towing passenger steamer CALEDONIAN PRINCESS (1769gt/1961) for conversion to a leisure/entertainments complex.   1984: Sold to Oil Transport Co S.A, Santo Domingo, Dominican Republic.   Poole registry closed.   Registered at Santo Domingo as **OTC ELIZABETH**.   1996: With only one engine operational and attempting to refloat a stranded coastal trading vessel, stranded on the east coast of the Dominican Republic.   Salvage unsuccessful and left as lies.

ROSS DARING 304785 GY707 Motor stern trawler	1488 25.05.1963 08.10.1963	201 72	85.0 23.0 12.5	Ruston Paxman 450bhp 8-cyl CP 11.0 knots	Ross Trawlers Ltd, Grimsby

First Cochrane-built stern trawler.   Cost to build £109,278.   Government grant £23,700.
08.10.1963: Registered at Grimsby (GY707).   1965: Suffered two engine "blow outs" in the space of 8 months.   24.02.1966: Alteration of particulars following fitting of 12-cylinder 520bhp oil engine by Davey, Paxman & Co Ltd, Colchester; 10 knots.   04.10.1966: Remeasured to 132grt, 68net.   06.1968: Sold to Productos Marinos S.A, Paita, Peru in an en bloc deal with ROSS DAINTY (see yard no.1498).
24.06.1968: Grimsby registry closed.   Registered at Paita as **AUDAZ**.   03.1980: Registered as a total loss.   Paita registry closed.

ROSS VALIANT 304793 IMO 6406555 GY729 Diesel-electric freezer stern trawler	1489 30.01.1964 20.07.1964	1156 528	190.0 36.5 16.75	Davey Paxman 1650bhp 3 x 8-cyl 13.5 knots	Ross Trawlers Ltd, Grimsby

Cost to build £513,099.   Government Grant £110,000.   First freezer stern trawler built for Grimsby and at the time of her launch the biggest ship built by Cochrane.
16.07.1964: Registered at Grimsby.   24.07.1964: First trip sailed for Labrador.   01.03.1968: Sold to Ross Freezer Trawlers Ltd, Grimsby.   01.07.1969: Became part of the British United Trawlers (B.U.T.) fleet.   Adopted B.U.T. livery.   Book value £299,559.   10.11.1970: Whilst fishing 200 miles off Tromsø went to the aid of the Hull trawler C S FORESTER (H86) (768grt/1969) which was disabled after a fire in the engine room.   Connected and delivered to Harstad.   30.09.1971: Sold to Northern Trawlers Ltd, Grimsby.   05.1975: Laid up for sale as uneconomic (fishroom too small).   09.1975: Sold to Skipper Gottfred Joensen P/R, Sandavág, Faroe Islands.   Remeasured to 803gt, 291net.   09.1975: Grimsby registry closed.   Registered at Sandavág as **V. U. HAMMERSHAIMB** (engaged in shrimp fishing at Greenland).
04.1978: Re-engined with three 8-cylinder oil engines by Klöckner-Humbolt-Deutz, Cologne, total 1160bhp.   1981: Sold to Kanajaq Trawl ApS, Godthåb, Greenland.   Sandavág registry closed.   Registered at Godthåb as **AVOQ**.   1989: Sold to Miteq Trawl ApS, managed by Måløy Verft A/S, Norway.   Registered at Godthåb as **NERPILIK**.   Remeasured to 1298gt, 389net.   1992: Sold to Korfskiy Rybozavod, Petropavlovsk / Kamchatskiy, Russia.   Godthåb registry closed.   Registered at Monrovia as **ZALIV KORFA**.   Re-engined with 8-cylinder oil engine by Wärtsilä-Wichmann A/S, Rubbestadneset, 3260bhp; 16 knots.   1996: Sold to A/O Koryakryba Joint Stock Co (A/O Koryakryba), Korf, Russia.   2010: Deleted from Lloyd's Register of Shipping - "continued existence in doubt".

Stella Altair (1485)

(Cochrane archive)

Workman (1487)

(Cochrane archive)

Ross Valiant (1489)

(Cochrane archive)

CAPE KENNEDY 305780 H353 D-E stern trawler	1490 12.06.1964 08.03.1965	1156 531	190.0 36.5 24.5	Mirrlees National 1950bhp 3 x 8-cyl 14.0 knots	Hudson Bros Trawlers Ltd Hull

Hudson Bothers first freezer trawler.   Cost to build £507,826.   Government Grant £110,000.
01.03.1965: Registered at Hull (H353).   12.04.1965: First trip sailed to Newfoundland grounds.   07.01.1966: Registered at Hull as **ROSS KENNEDY** (H353). Ross Group livery adopted.   11.05.1966: Registered at Hull as **ROSS INTREPID** (H353).   01.03.1968: Sold to Ross Freezer Trawlers Ltd, Grimsby.   01.07.1969: Became part of the British United Trawlers (B.U.T.) fleet. Adopted B.U.T livery.
30.09.1971: Sold to Northern Trawlers Ltd, Grimsby.   10.1973: Transferred to fish out of Grimsby (B.U.T. fleet re-organisation).
04.1975: Sold to Tor & Harald Ostervold, Bergen, Norway.   24.04.1975: Hull registry closed.   Registered at Bergen as **MALENE OSTERVOLD** (IMO 6415051).   1978: Sold to Harald Ostervold, Bergen.   Re-engined with 8-cylinder 3997bhp oil engine by Mak Maschinenbau GmbH, Kiel.   Converted to a research vessel and remeasured to 1118gt, 421net.   1989: Sold to Malene Ostervold Shipping A/S, Torangsvåg.   2000: Sold to Harald Ostervold, Torangsvåg.   16.02.2001: Sold to Norfield Shipping A/S, Torangsvåg.
07.2013: Bergen registry closed.   Registered at Nassau, Bahamas.   2015: Still registered.

ROSS DELIGHT 304791 GY727 Motor stern trawler	1491 21.08.1963 27.01.1964	201 72	85.0 23.0 12.5	Davey Paxman 450bhp 8-cyl 11.2 knots	Ross Trawlers Ltd, Grimsby

Cost to build £111,214 (WFA grant £23,700).
28.01.1964: Registered at Grimsby (GY727).   12.06.1968: Sold to Offshore Charters Ltd, Grimsby.   15.11.1968: Registered at Grimsby as **OFFSHORE DELIGHT**.   27.11.1968: Remeasured to 182grt, 68net.   26.11.1971: Sold to The National Environment Research Council, Cardiff.   17.01.1972: Grimsby registry closed. Converted to a research vessel.   Registered at Cardiff as **EDWARD FORBES**.   1979: Sold to Bristol Channel Divers Ltd, Cardiff.   Registered at Cardiff as **DONQUESTO**.   01.1984: Sold to Armafin S.T.L. Societa Finanziaria ~ Di Armamento, Rome, Italy. Cardiff registry closed. Converted to an oceanographic research vessel. Registered at Rome as **CIPREA**.
1991: Major refit and re-engined with a 12-cylinder 700bhp oil engine by Gutierrez Ascunce Corp S.A. Zumaya.   08.1996: Sold to G.E.O. Mare S.t.L, Rome.   01.1998: Sold to Italian owners.   Renamed **DON QUESTO** (IMO 5422253). Fitted out for diving/live aboard and based in Sudan.   2015: Believed still in service.

J A LAMEY 303897 IMO 5415949 Motor tug	1492 08.07.1963 14.03.1964	216	96.0 26.6	Motorenwerk 1360bhp 8-cyl 12.0 knots	J H Lamey Ltd, Liverpool

03.1964: Registered at Liverpool.   07.1968: Company and assets sold to Alexandra Towing Co Ltd, Liverpool.   1970: Registered at Liverpool as **HORNBY**.   1984: Sold to John McLoughlin & Son (Shipping) Ltd, Larne, Co Antrim.   Registered at Liverpool as **SAMUEL F**.
04.2007: Sold to Mediterranean buyers reportedly for use in a Malta-based fish farm project.   20.05.2007: Sailed Howth, Co Dublin, for Malta.   2015: Believed still in service.

JÖRUNDUR II IMO 5424380 RE299 Motor trawler	1493 19.09.1963 02.03.1964	267 106	104.7 25.0 12.0	Blackstone 800bhp 8-cyl CPP 12.5 knots	Jörundur H/f. Reykjavik, Iceland

Combination fishing vessel Seine/trawl/longlines
01.04.1963: Registered at Reykjavik (RE299).   24.06.1969: Sold to Jökull H/F, Raufarhöfn.   Reykjavik registry closed.   Registered Raufarhöfn as **JÖKULL**.   30.03.1973: Sold to Útnes H/F, Rif.   Raufarhöfn registry closed.   Registered at Rif as **HAMAR** (SH224).
05.1974: Remeasured to 233grt, 82net.   26.10.1974: Sold to Kristinn J Friðþjófsson, Rif.   1976: Re-engined with 1000bhp oil engine by Lister Blackstone.   1979: Rebuilt with deck fully enclosed; remeasured to 235grt.   2000: Stern rebuilt and lengthened to 112.6 ft (34,32m) by Thorgeir og Ellert yard, Akranes, Iceland.   Remeasured to 244gt, 103net.   2015: Still in service.

JÖRUNDUR III IMO 5427710 RE300 Motor trawler	1494 19.10.1963 15.04.1964	267 104	104.0 25.0 12.0	Blackstone 800bhp 8-cyl CPP 12.5 knots	Jörundur H/f. Reykjavik, Iceland

Combination fishing vessel Seine/trawl/longlines
06.05.1963: Registered at Reykjavik (RE300).   01.12.1972: Sold to Richard Sighvatsson & Sigurður G. Ásbjörnsson, Vestmannaeyjar.   Reykjavik registry closed.   10.01.1973: Registered at Vestmannaeyjar as **HÁSTEINN** (VE355).   01.02.1973: Registered at Vestmannaeyjar as **ÁSVER** (VE355).   05.1974: Remeasured to 233grt, 82net.   23.12.1976: Sold to Jakob Sigurðsson & Magnús Grímsson, Reykjavík.   Registered at Reykjavik as **SÆBORG** (RE20).   22.05.1980: Sold to Sjófang H/F, Reykjavik.   1982: Re-engined with 800bhp oil engine by Mirrlees Blackstone.   27.10.1992: Sold to Visir H/F, Grindavik.   Reykjavik registry closed.   Registered at Grindavik (GK457).   15.06.1996: Fishing as a purse seiner; eleven crew.   In good weather conditions returning from the fishing grounds with 3,300 kits (200 tons) of herring when some 165 nautical miles east of Iceland started to take in water forward.   Mayday sent and crew managed to put three liferafts in the water and abandoned the ship shortly before she started to settle by the head exposing her propeller and rudder before foundering. Purse seiner EÐVALDS (SF20) (336grt/1980) bound for the fishing grounds picked up all the crew.   Grindavik registry closed.

SJOLLEN IMO 6411421 Motor tug, ice class 1	1495 14.04.1964 15.10.1964	270 80	91.0 28.25 16.75	Ruston & Hornsby 2100bhp 2 x 8-cyl CPP	Malmö Bogser A/B Malmö, Sweden

10.1964: Registered at Malmö.   1977: Sold to Broströms Rederi A/B (Röda Bolaget), Malmö.   1983: Sold to Pohle Slepe Bergningstjeneste, Tromsø.   Malmö registry closed.   Registered at Tromsø as **LØVE**.   11.1991: Port engine replaced with 8-cylinder 1050bhp oil engine by Ruston Diesels.   01.1994: Sold to Neptun Rederi, Tromsø.   01.1994: Registered at Tromsø as **BRYTEREN**.   Remeasured to 256gt, 76net.
2000: Sold to Morland & Karlsen A/S, Arendal.   09.2000: Tromsø registry closed.   Registered at Arendal as **TUG FRISØY**.   01.2004: Sold to K Jousmaa Ky, Tammisaari, Finland.   Arendal registry closed.   Registered at Tammisaari as **POLARIS**.   21.05.2007: Sold to Rauma Chartering & Towage Agency OY AB, Ltd, Rauma, Finland.   06.2012: Tammisaari registry closed.   Registered at Rauma.   2015: Still in service.

The **Ross Delight** (1491)
arriving at Grimsby for the
first time.

(Jonathan Grobler
collection)

The **Hornby** (1492) with the
familiar Liverpool background.
The tug was originally named
J A Lamey.

(Stuart Emery collection)

The **Sjollen** (1495) was the
first of several tugs built for
Swedish owners.

(Cochrane archive)

KARL IMO 6420642 TM tug, ice class 1	1496 25.08.1964 19.01.1965	269 81	91.0 28.25 16.75	Ruston & Hornsby 3024bhp 2 x 6-cyl CPP	Göteborgs Bogserings & Bärgnings, Gothenburg, Sweden

01.1965: Delivered to Göteborgs Bogserings & Bärgnings AB, Gothenburg, as **KARL**. Registered at Gothenburg. 1977: Owners restyled Broströms Rederi AB, Gothenburg and in 1984 to AB Neptun-Röda Bolaget, Gothenburg, which later became Röda Bolaget AB, Gothenburg. 12.2000: Sold to Vänerhamn A/B, Karlstad. Gothenburg registry closed. Registered at Karlstad as **KARL AF KARLSTAD**. 2015: Still in service.

ENGLISHMAN 305784 TM tug	1497 21.11.1964 27.04.1965	574 30	130.0 33.0 19.8	Ruston & Hornsby 2930bhp 2 x 6-cyl 15.0 knots	United Towing Co Ltd, Hull

30.04.1965: Registered at Hull. 30.01.1970: Company restyled United Towing Ltd. 03.07.1973: Sold to United Towing (Englishman) Ltd, Hull. 1973: Used extensively as a defence tug during the second 'Cod War' with Iceland. 22.12.1980: Sold to Societa Armamento Gestione Navi Agenzia Marittima S.r.L (S. AR.GE.NAVI), Naples, Italy. 23.01.1981: Hull registry closed. Registered at Naples as **JUMBO PRIMO**. 06.1994: Sold to T.S.A. Tugs Ltd, Westcliffe-on-Sea, Essex. Naples registry closed. Registered at Kingstown, St. Vincent & the Grenadines, as **TOWING WITCH**. 04.06.1997: Off the Azores towing two pontoons barges. In approximate position 33.28N 33.10W had engine room fire and suffered extensive damage. Crew taken off by the motor vessel DOLE EUROPA (10584gt/1994). Salvage tug FOTTY KRYLOV (5250gt/1997) connected to tug and barges and commenced tow to Azores. 10.06.1997: Delivered Ponto Delgado. Survey revealed extensive damage and declared a constructive total loss (CTL). Sold to Spanish shipbreakers broken up by Viguesa de Chatarra SL, Vigo. Registry closed.

ROSS DAINTY 304798 GY1355 Motor stern trawler	1498 19.01.1965 25.05.1965	202 68	85.25 23.0 12.5	Davey Paxman 450bhp 8-cyl CPP 12.2 knots	Ross Trawlers Ltd, Grimsby

Cost to build £112,436 (WFA Grant £23,700).
24.05.1965: Registered at Grimsby (GY1355). 06.1968: Sold to Productos Marinos S.A, Peru, in an en bloc deal with ROSS DARING (see yard no.1488). 24.06.1968: Grimsby registry closed. Registered at Paita as **INTREPIDO**. 1982: Sold to Pescamar S.r.L, Paita, Peru. 1988: Sold to Ribar S.A, Peru. Registered at Paita as **RIBAR V**. 19.08.1988: Foundered approximately 34 miles NW of Chimbote, Peru. Paita registry closed.

HERAKLES IMO 6518140 Oceangoing motor tug Ice strengthened	1499 01.07.1965 15.12.1965	492 145	127.5 33.5 16.25	Ruston & Hornsby 4752bhp 2 x 9-cyl CPP 14.0 knots	Bergnings & Dykeri A/B Neptun, Stockholm, Sweden

Ice strengthened. 12.1965: Registered at Stockholm. 1968: Sold to Göteborgs Bogserings & Bärgnings A/S, Gothenburg. Stockholm registry closed. Registered at Gothenburg as **DAN**. 1974: Sold to Ministry of Defence (Navy), London, for a sum believed to be £685,000. Gothenburg registry closed. 04.1974: Commissioned as HMS **WAKEFUL** (P.No.A236). Based HMS NEPTUNE, Faslane Naval Base, as a submarine escort ship and exercise target. As required also assigned to fishery protection and miscellaneous naval duties. 15.10.1987: Arrived Portsmouth. 30.10.1987: De-commissioned. Offered for sale. 05.1988: Sold to Megalohari Hellenic Tugboats, Piræus, Greece. 06.05.1988: Sailed Portsmouth for Greece. Refitted as a tug. Remeasured to 483gt, 145net. Registered at Piræus as **AEGEON PELAGOS**. Later registered at Piræus as **AEGEAN PELAGOS**. 2015: Still in service.

ROSS FAME 307545 IMO 6510813 GY1360 Motor stern trawler	1500 19.03.1965 17.08.1965	457 183	120.0 30.0 19.5	Davey Paxman 950bhp 12-cyl CPP 12.0 knots	Ross Trawlers Ltd, Grimsby

Cost to build £216,336 (WFA Grant £42,000).
06.08.1965: Registered at Grimsby (GY1360). 28.11.1968: Alteration of particulars after re-engined with an 8-cylinder 1052bhp oil engine by English Electric Co Ltd, Newton-Le-Willows. Remeasured to 320grt, 103 net. 01.07.1969: Became part of the British United Trawlers (B.U.T.) fleet. B.U.T. livery adopted. Book value £140,758. 21.01.1971: Sold to Bonavista Cold Storage Co Ltd, St John's, Newfoundland, Canada. 11.11.1971: Grimsby registry closed. Registered at St. John's as **FAME V**. 1984: Sold to The Lake Group Ltd, St. John's. 1986: Sold to Fishery Products International Ltd, St John's. 10.1990: St John's registry closed. Registry at Belize. 1993: Registered at Belize as **FURY**. 09.1993: Sold to Becker Broadcast Systems Corp, Belize. Registered at Belize as **BOLERO EXPRESS**. 1996: Sold to Trans American Holdings Ltd, Belize. Converted to dry cargo vessel. Remeasured to 467gt, 140net. Registered at Belize as **CARIBBEAN SEAHORSE**. 08.1999: Sold to International Ship Co, North Miami, Florida. Belize registry closed. 08.1999: Registered at San Lorenzo, Honduras. 2015: Still in service.

ROSS FORTUNE 307549 IMO 6515801 GY1365 Motor stern trawler	1501 15.04.1965 28.10.1965	457 183	120.0 30.0 19.5	Davey Paxman 950bhp 12-cyl CPP 12.0 knots	Ross Trawlers Ltd, Grimsby

Cost to build £212,909 (WFA Grant £42,000).
15.10.1965: Registered at Grimsby (GY1365). 11.12.1968: Alteration of particulars after re-engined with an 8-cylinder 1052bhp oil engine by English Electric Co Ltd, Newton-le-Willows. Remeasured to 320grt, 103net. 01.07.1969: Became part of the British United Trawlers (B.U.T.) fleet. B.U.T. livery adopted. Book value £138,124. 06.03.1970: Sold to Wyre Trawlers Ltd, Fleetwood for £159,528. 06.07.1970: Grimsby registry closed. 07.07.1970: Registered at Fleetwood (FD281). 04.06.1971: Sold to Bonavista Cold Storage Co Ltd, St John's, Newfoundland, Canada. 26.01.1972: Fleetwood registry closed. Registered at St.John's as **FURY V**. 02.1972: Re-engined with a 12-cylinder 1500bhp oil engine by General Motors Corp, La Grange, GA. 1984: Sold to The Lake Group Ltd, St John's. 1986: Sold to Fishery Products International Ltd, St John's. 1993: Sold to Naviera Fury S de R.L, Tegucigalpa, Honduras. St John's registry closed. Registered at San Lorenzo. 2002: Registered at San Lorenzo as **FURY**. 2002: Removed from Lloyd's Register of Shipping - "Vessel's continued existence in doubt".

Not built	1502 Order cancelled				

ROSS VANGUARD 307561 GY1372 Motor stern trawler freezer	1503 12.10.1965 17.06.1966	1488 643	200.0 30.0 19.5	Ruston & Hornsby 2150bhp 9-cyl 13.5 knots	Ross Trawlers Ltd, Grimsby

Cost to build £512,580.
15.06.1966: Registered at Grimsby (GY1372). 24.07.1964: First trip sailed to Labrador grounds. 01.07.1969: Became part of the British United Trawlers (B.U.T.) fleet. B.U.T. livery adopted. Book value £438,721. 01.07.1971: Remeasured to 1060gt, 354net.
02.05.1978: Transferred to fish out of Hull. 1977-1978: B.U.T. transferred all their freezer trawler fleet to Hull. 12.08.1978: Fishing off Bear Island (Sk Peter Costello). The Grimsby freezer trawler ROMAN (GY253) (Sk Terry Bascombe) was in company when at 10am a serious fire broke out in her engine room, three factory hands who went in to fight the fire were overcome by fumes and despite repeated attempts by officers wearing breathing apparatus to reach them they were lost. Responded to distress call and attempted to put fire hoses on board, but these parted because ROMAN had no steerage control. ROMAN's sister ship, GOTH (GY252) (Sk Jock Kerr) also responded and after ten survivors had been transferred, connected and with GOTH commenced tow to Honningsvåg. 27.07.1979: Landed at Hull and put up for sale. 31.10.1979: Sold to Wale Sea Foods Ltd, Lagos, Nigeria. Renamed **ARAMOKO**. Company did not register their title.
Mid-1980s: Reported as subject to piracy and gutted off Apapa, Lagos. 14.10.1986: Papers filed for possession. 25.04.1989: Removed from Lloyd's Register of Shipping - "Vessel's continued existence in doubt". 11.05.1989: Grimsby registry closed.

ROSS ILLUSTRIOUS 308560 IMO 6610039 H419 Motor stern trawler freezer	1504 05.04.1966 23.09.1966	1488 629	200.0 39.5 26.0	Ruston & Hornsby 2150bhp 9-cyl 13.5 knots	Hudson Brothers (Trawlers) Ltd, Hull

Cost to build £537,720.
20.09.1966: Registered at Hull (H419). 29.09.1966: First trip sailed for Newfoundland, Grand Banks. 01.07.1969: Became part of the British United Trawlers (B.U.T.) fleet. B.U.T. livery adopted. Book value £462,585. 18.01.1971: Remeasured to 1076gt, 354net.
10.1973: Transferred to fish out of Grimsby. 16.11.1973: Sold to British United Trawlers, Grimsby. 30.06.1978: B.U.T. transferred all their freezer trawler fleet to Hull. 18.02.1980: Laid up; fishing registration closed. 1982: Sold to Seaboard Offshore Ltd, Tain, Ross-shire.
07.1983: Sold to Seaboard Offshore (Diving & Pollution) Ltd, Tain, Ross-shire Converted to a diving support ship. 29.07.1983: Hull registry closed. Registered at Inverness as **SEABOARD ILLUSTRIOUS**. 1987: Converted to an offshore platform standby safety vessel.
1995: Sold to Hornbeck Offshore Ltd, Inverness. Registered at Inverness as **HORNBECK ILLUSTRIOUS**. Remeasured to 1433grt, 429net.
05.1998: Sold to Tidewater Marine IOM Ltd, Douglas, Isle of Man. Inverness registry closed. Registered at Port Vila,Vanuatu, as **CEANIC ROVER**. 1999: Sold to BUE North Sea Ltd, Leith. 09.1999: Sold to Ceanic Worldwide Ltd, Panama. Port Vila registry closed. Registered at Panama as **SEAWAY ROVER**. 05. 2003: Sold to Acergy Shipping Ltd, Westhill, Aberdeenshire. 01.2005: Sold to Hallström Holdings Oceanics AB, Yxlan, Sweden. 04. 2007: Sold to Seaway Rover Pte Ltd, Singapore. 16.01.2008: Sold to Peacegate Oil & Gas Ltd, Victoria Island, Lagos, Nigeria. Panama registry closed. 01.2008: Registered at Lagos as **SEAWAY AGBAMU**. 2015: Still in service.

EAST SHORE 308459 TM supply ship	1505 25.11.1965 04.04.1966	669 232	148.0 37.5 15.5	Blackstone 1600bhp 2 x 8-cyl 12.0 knots	London & Rochester Trading Co Ltd, Rochester

Contract signed £175,000.
04.04.1966: Completed for Offshore Marine Ltd, Rochester. Registered at Rochester. 1968: Company and vessels sold to Cunard Group, London. 16.01.1974: Sailed from Sète, France (Capt Roger J Parham); six crew total, with 249 tons deck cargo to attend the drill ship PELICAN. 17.01.1974: In the early hours a mistral developed from NW reaching gale force at daybreak. Marseilles Radio received a Mayday message at 0750 GMT stating cargo had shifted and immediate assistance required. Position was 80 miles from land. More messages were received by Marseilles Radio but doubtful if ship received the replies. At 1126 GMT motor vessel FELICE d'AMICO (8522gt) arrived on scene and made unsuccessful attempts to establish a connection. In the early evening the FELICE d'AMICO's deck cargo shifted and she had to look to herself but at 2052 GMT when some two miles away received a message reporting that the crew were abandoning ship after list increased. Foundered in position 42.06N 04.47E and despite a search by six ships and aircraft no survivors found. Registry closed.

WELSHMAN 308559 IMO 6613457 TM tug	1506 24.05.1966 07.11.1966	451	112.0 31.0 15.5	Ruston & Hornsby 2820bhp 2 x 6-cyl 13.5 knots	United Towing Co Ltd, Hull

04.11.1966: Registered at Hull. 30.01.1970: Company restyled United Towing Ltd. 30.04.1973: Sold to United Towing (Welshman) Ltd. Hull. 06.10.1978: Sold to J P Knight Ltd, Rochester. 27.11.1978: Hull registry closed. Registered at Inverness as **KINLUCE**. 1980: Sold to Barclays Mercantile Industrial Finance Ltd, London. 09.1988: Sold to Klyne Tugs (Lowestoft) Ltd, Lowestoft. Inverness registry closed. Registered at Lowestoft as **ANGLIAN LADY II**. 1992: Sold to Remolques del Atlantico Remay S, La Coruña, Spain. Lowestoft registry closed. Registered at La Coruña as **VIKINGO**. 1992: Sold to Puestas y Obras SA, La Coruña. Registered at La Coruña as **KOCHAB**. 1997: Sold to D.D.Y. de Comercio Exterior SA, Spain. 03.1998: Sold to Keypana Naviera SA, Panama. La Coruña registry closed. Registered at Panama. Remeasured to 536gt, 161net - 11knots. 11. 2001: Sold to Gold Star Marine Management Corp, Panama. 2007: Sold to United States shipbreakers. 26.01.2007: Arrived Rockport, Texas, for breaking up. Registry closed

IRISHMAN 308569 IMO 6617128 TM tug	1507 18.07.1966 11.01.1967	451	112.0 31.0 15.5	Ruston & Hornsby 2820bhp 2 x 6-cyl 13.5 knots	United Towing Co Ltd, Hull

10.01.1967: Registered at Hull. 30.01.1970: Company restyled United Towing Ltd. 30.04.1973: Sold to United Towing (Irishman) Ltd, Hull. 1973: Used extensively as a defence tug during the second 'Cod War' with Iceland. 01.04.1974: Sold to Star Offshore Services (Tugs) Ltd. Hull. 09.08.1976: Sold to Guybulk Shipping Ltd, Hamilton, Bermuda. 12.08.1976: Hull registry closed. Registered at Hamilton as **KWAKWANI**. 1978: Sold to International Towage Ltd, Ramsgate. Registered at Hamilton as **LORNA B**. 1980: Sold to I.T.L. International Towing Ltd, Ramsgate. 09.1981: Sold to Pacific Standard Towing Ltd, Vancouver, B.C., Canada. Hamilton registry closed. Registered at London as **PACIFIC STANDARD**. 1985: London registry closed. Registered at Vancouver, BC. 2000: Sold to Mckeil Work Boats Ltd, Hamilton, Ontario. Fitted as a pusher tug. 2012: For sale. 2015: Still in service.

Ross Dainty (1498)

(Cochrane archive)

Ross Vanguard (1503)

(Jonathan Grobler
collection)

The **Welshman** (1506)
outward bound from Great
Yarmouth in August 1968.

(Des Harris,
Andrew Wiltshire collection)

PRONTO 309916 IMO 6620022 Motor oil tanker	1508 16.09.1966 25.08.1967	588 243	162.6 34.0 15.5	Mirrlees National 708bhp 6-cyl 9.5 knots	Shell Mex & BP Ltd, London

Classed for use in the Thames or Mersey estuary service.
08.1967: Registered at London. 1970: Lengthened to 184.5 ft - 652grt, 310net, 1008dwt. Classed as a tank barge. 1976: Sold to BP Oil Ltd, London, on de-merging of Shell Mex & BP Ltd. Registered at London as **BP ALERT**. 1983: Sold to Songhai Petroleum Co Ltd, Apapa, Lagos, Nigeria. Registered at London as **SONGHAI No.1**. Classed for Nigerian coastal service as an oil products tanker. 1985: Registered at Lagos, Nigeria. 2015: Still in service.

PERFECTO 334609 Motor oil tanker	1509 20.01.1967 24.08.1967	588 243	162.6 34.0 15.5	Mirrlees National 708bhp 6-cyl 9.5 knots	Shell Mex & BP Ltd, London

Classed for Thames or Mersey estuary service.
08.1967: Registered at London. 1969: Lengthened to 184.5 ft - 652grt, 310net, 1008dwt. 09.1975: Re-engined with 8-cylinder 580bhp oil engine by Alpha-Diesel A/S, Frederikshavn - 9 knots. 1979: Sold to Shell U.K. Ltd, London, on de-merging of Shell Mex & BP Ltd. Registered at London as **SHELL DRIVER**. 1989: Sold to C Crawley Ltd, Gravesend. Registered at London as **PERFECTO**. 1997: Sold to Nortech (Scotland) Ltd, Coatbridge, Lanarkshire, Scotland. 10.2000: Registered at São Tomé, São Tomé & Príncipe. 12.2000: Registered at São Tomé as **FECTO** (IMO 6707985). 01.2001: Sold to Petrostar Nigeria Ltd, Apapa, Lagos, Nigeria. 2015: Believed still in service.

WEST SHORE 333518 TM supply ship	1510 15.11.1966 06.03.1967	698 245	152.0 37.6 15.5	Blackstone 1600bhp 2 x 8-cyl 12.0 knots	Offshore Marine Ltd, Rochester

03.1967: Registered at Rochester. 1968: Company and vessels sold to Cunard Group, London. 07.01.1971: Foundered near drilling platform OCEAN VIKING in position 56.16N 03.09E after engine room flooded during a blizzard. All thirteen crew saved. Registry closed.

KENT SHORE 333523 IMO 6709880 TM supply ship	1511 28.02.1967 16.05.1967	664 241	152.0 37.6 15.5	Blackstone 1600bhp 2 x 8-cyl 12.0 knots	Offshore Marine Ltd, Rochester

05.1967: Registered at Rochester. 1968: Company and vessels sold to Cunard Group, London. 1980: Company and vessels sold to Zapata Gulf Marine Corp, Houston, Texas. Registered at Rochester as **KENT SERVICE**. 1984: Sold to Deventel Ltd c/o Nomis Ltd, Aberdeen. Rochester registry closed. Registered at Aberdeen as **VICTORIA KENT**. 02.08.1985: Sold to Putford Enterprises Ltd, Lowestoft. Aberdeen registry closed. Registered at Lowestoft as **PUTFORD TERN**. 1993: Converted to offshore platform standby safety vessel. 02.2000: Sold to George Gantt, USA. Lowestoft registry closed. Registered at Panama as **LYNN B**. 05.2000: Sold to Oaklea Marine Ltd, Gibraltar. Registered at Panama as **CAPE ENDURANCE**. 11.06.2003: Sold to Wellness International Network Ltd, Gibraltar. 2015: Believed still in service.

SUFFOLK SHORE 333527 IMO 6714847 TM supply ship	1512 25.04.1967 28.06.1967	664 241	152.0 37.6 15.5	Blackstone 1600bhp 2 x 8-cyl 12.0 knots	Offshore Marine Ltd, Rochester

05.1967: Registered at Rochester. 1968: Company and vessels sold to Cunard Group, London. 1980: Company and vessels sold to Zapata Gulf Marine Corp, Houston, Texas. Registered at Rochester as **SUFFOLK SERVICE**. 1984: Sold to Putford Enterprises Ltd, Lowestoft. 1985: Converted to diving support ship. Rochester registry closed. Registered at Lowestoft as **PUTFORD SKUA**. 04.1991: Converted to offshore platform standby safety vessel. 03.2004: Sold to Echoscan Ltd, Leeds, Yorkshire. Converted to a research vessel. Lowestoft registry closed. Registered at Panama as **SCAN WARRIOR**. 08.2006: Converted to offshore support vessel with diving facilities. 07.2007: Sold to Asian Navigation Ltd, Hong Kong. Registered at Panama as **ASIAN WARRIOR**. 03.2008: Panama registry closed. Registered Basseterre, St Kitts & Nevis. 2015: Still in service.

Not built	1513 Order cancelled				

AXEL IMO 6720743 Motor fifi tug, ice class 1	1514 21.06.1967 26.09.1967	220	85.0 27.8 15.0	Ruston & Hornsby 2460bhp 9-cyl CPP 13.5 knots	Göteborgs Bogserings & Bärgnings, Gothenburg, Sweden

09.1967: Registered at Stockholm. 1977: Following merger of Röda Bolaget/Neptunbolaget/Broström, sold to Röda Bolaget - Broströms Rederi A/B, Gothenburg. 1983: Owners restyled A/B Neptun - Röda Bolaget, Gothenburg. 1986: Sold to Scandinavian Towage & Salvage (Scantugs) A/B, Gothenburg, for the sum of 4.5m kr. 1987: Sold to Halmstads kommuns Hamnförvaltning, Halmstad, for the sum of 5.0m kr. Stockholm registry closed. Registered at Halmstad as **AXEL AV HALMSTAD**. 1989: Sold to Halmstads Hamn Ab, Halmstad, for the sum of 4.4m kr. 09.1991: Company restyled Halmstads Stuveri AB, Halmstad. 03.1992: Sold to Röda Bolaget A/B, Gothenburg' Halmstad registry closed. Registered at Gothenburg as **AXEL AF GÖTEBORG**. 27.01.2001: Sold to Rundviks Rederi A/B, Nordmaling. 06.2013: Registered at Rundvik as **AXEL AF RUNDVIK**. 2015: Still in service.

UTTERN IMO 6721345 Motor fifi tug, ice class 1	1515 21.07.1967 10.11.1967	220	85.0 27.8 15.0	Ruston & Hornsby 2460bhp 9-cyl CPP 13.5 knots	Malmö Bogserings AB Malmö, Sweden

11.1967: Registered at Malmö. 1968: Sold to Göteborgs Bogserings & Bärgnings A/B, Gothenburg. Malmo registry closed. Registered at Gothenburg as **HARALD**. 1977: Ownership transferred to Broströms Rederi AB – Röda Bolaget,. 1984: Ownership transferred to AB Neptun-Röda Bolaget. 1987: Sold to Scandinavian Towage & Salvage (Scantugs) A/B, Gothenburg. 1988: Sold to Ykspihlaja Hinaus-Bogsering, Kokkola, Finland, for 4.5m kr. Gothenburg registry closed. Registered at Kokkola. 1992: Owners restyled Perämeren Hinaus Oy. 08.1994: Sold to Halmstads Stuveri AB, Halmstad, Sweden, for 5.4m kr. Kokkola registry closed. Registered at Halmstad as **STIG**. 08.2009: Sold to Oskarhamns Hamn AB, Oskarshamn. 08.2009: Halmstad registry closed. Registered at Oskarshamn as **OSCAR AF OSKARSHAMN** (written on ship as **OSCAR AF OSCARSHAMN**). 2015: Still in service.

GÖSTA IMO 6800438 Motor fifi tug, ice class 1	1516 01.11.1967 30.01.1968	220	85.1 27.6 15.0	Ruston & Hornsby 2460bhp 9-cyl CPP 12.5 knots	Göteborgs Bogserings & Bärgnings, Gothenburg, Sweden	

01.1968: Registered at Gothenburg.   1977: Sold to Broströms Rederi A/B, Gothenburg.   1977: Following merger of Röda Bolaget/Neptunbolaget/Broströms, sold to AB Neptun - Röda Bolaget, Gothenburg.   1997: Transferred to work at Norrköping. 1997: Gothenburg registry closed. Registered at Norrköping. 12. 2003: Sold to Towage & Marine Assistance UAB, Klaipeda, Lithuania. Gothenburg registry closed. Registered at Klaipeda as **TAK-8**.   01.2004: Sold to OOO 'Grifon', St. Petersburg, Russia. Klaipeda registry closed. Registered at St Petersburg and renamed **GRIFON 3**.   04.2004: Sold to Svitzer Sverige AB, Copenhagen.   05.2006: Sold to Nord-West Towage, St Petersburg. Registered at St Petersburg as **TAK-8**.   2008: Sold to Towage & Marine Assistance, Klaipeda, Lithuania. St Petersburg registry closed. Registered at Klaipeda.   23.05.2011: Sold to Liepaja Special Economic Zone, Liepaja, Latvia. 05.2011: Klaipeda registry closed. Registered at Liepaja.   2015: Still in service.

FREDERIC B INGRAM TM tug	1517 20.10.1967 26.01.1968	387	112.0 31.0 15.5	Ruston & Hornsby 4200bhp 2 x 8-cyl 13.0 knots	Ingram Marine Ltd, Harvey, Louisiana, USA

01.68: Registered at Panama.   1972: Sold to Ingram Contractors S.A, Harvey, La, USA. Registered at Panama as **JARAMAC XXVIII**. 25.07.1973: Sold to United Towing Ltd Hull. Modified for anchor handling for North Sea oil industry.   10.1973: Panama registry closed. 08.10.1973: Registered at Hull to United Towing (Norman) Ltd, Hull, as **NORMAN** Official No.359193.   04.12.1973: Alteration to engine particulars, uprated to 5000 bhp.   01.04.1974: Sold to Star Offshore Services (Tugs) Ltd, London.   13.12.1975: In heavy seas off the Humber capsized and foundered 38 miles off Spurn Point in position 53.24N 01.01E following engine room flooding. All crew saved. 08.06.1976: Register closed.

ROSS IMPLACABLE 334090 IMO 6812431 H6 Motor stern trawler freezer	1518 16.03.1968 27.07.1968	1042 510	200.0 39.6	Ruston & Hornsby 2150bhp 9-cyl 14.5 knots	Ross Freezer Trawlers Ltd, Hull

Cost to build £523,219 (40 per cent Government grant).
19.07.1968: Registered at Hull (H6).   02.08.1968: First trip sailed for White Sea grounds.   01.07.1969: Became part of the British United Trawlers (B.U.T.) fleet. B.U.T. livery adopted. Book value £357,278.   1969: B.U.T.'s top Hull trawler: 3,781 tons, 284 days at sea. 30.09.1971: Sold to Hellyer Bros Ltd, Hull.   04.02.1972: After engine failure, towed into Bergen by Grimsby trawler LORD BEATTY (GY11). 10.1973: Transferred to fish out of Grimsby.   10.12.1974: Towed to the Humber from Spitzbergen after engine failure.   05.1978: B.U.T. transferred all their freezer trawler fleet to Hull.   09.02.1979: Sold to Wilemace Ltd, Hull.   04.1982: Sold to Sherkat Taavoni Salyadi Mahdi, Teheran, Iran, for fishing in the Indian Ocean   20.04.1982: Hull registry closed.   04.1982: Registered at Teheran as **HAMOOR 1**. 1994: Sold to unlisted owners and registered in United Arab Emirates. Renamed **NICE TRAWLER**.   03.1995: Sold to Oman Fish Fortune, Port Sultan Qaboos, Oman. Registered at Port Sultan Qaboos as **AL MAUSUFAH**.   05.1996: Sold to Neptune Fishing Co Ltd,   Port Sultan Qaboos registry closed. Registered at Kingstown, St. Vincent & the Grenadines as **SEALION ONE**. Remeasured to 2359gt, 1415net 09.1998: Kingstown registry closed. Registered at Belize as **MARLIN**.   2000: Sold to Indian shipbreakers.   07.2000: Arrived Alang for breaking up. Registry closed.

RUDDERMAN 336939 IMO 6818552 Motor tanker	1519 12.06.1968 01.11.1968	1592 926	260.0 41.0 20.0	Drypool Brons 1500bhp 16-cyl 12.0 knots	Helmsman Shipping Co Ltd, London

11.1968: Registered at London.   1972: Owners restyled C Rowbotham & Sons (Management) Ltd, London.   1981: Sold to C Rowbotham Tankships Ltd, London.   1981: Sold to Ingram Overseas Ltd, London.   1985: Sold to Rowbotham Tankships Ltd, London.   1986: Sold to Societe Sofati & Soconav Ltee, Montreal, Canada. London registry closed. Registered at Montreal as **HENRI TELLIER**. 05.1995: Transferred to Port of Spain, Trinidad. Montreal registry closed. Registered at Port of Spain, Trinidad, as **RAMA J**. 01.1996: Converted to a water carrier. Remeasured to 1675gt, 1005net, 3004dwt.   1996: Sold to Epxon Oil Producing Co Ltd, Nigeria. Port of Spain registry closed.   01.1997: Registered at Lagos as **CHRISTINE**.   10.2005: Sold to shipbreakers and broken up. Registry closed

E BRONSON INGRAM IMO 6826119 TM tug	1520 20.08.1968 19.01.1969	381	112.0 31.0 15.5	Ruston & Hornsby 4200bhp 2 x 8-cyl 13.0 knots	Ingram Marine Ltd, Harvey, Louisiana, USA

01.1969: Registered at Panama.   1972: Sold to Oceanic Contractors Inc, Harvey, La, USA. Registered at Panama as **JARAMAC 42**. 21.02.1973: Sold to United Towing Ltd, Hull. Modified for anchor handling for North Sea oil industry.   03.1973: Panama registry closed. 09.03.1973: Registered at Hull as **SCOTSMAN**. Official No.359165. Remeasured to 412grt, 0net.   20.07.1973: Sold to United Towing (Scotsman) Ltd, Hull.   10.04.1978: Company restyled United Towing (Sandwich) Ltd, Hull.   07.1981: Sold to Arabian Bulk Trade Ltd Dammam, Saudi Arabia.   10.07.1981: Hull registry closed. Registered at Dammam as **AL BATTAL**.   1986: Sold to Great Lakes International Towing & Salvage Co Inc, Burlington, Ontario. Dammam registry closed. Registered at Hamilton, Bermuda, as **PETITE FORTE**. Fitted as a pusher tug. Remeasured to 369gt, 17net.   1991: Hamilton, Bermuda registry closed. Registered at Hamilton, Ontario. 02.10.2006: Downbound in Lake Erie pushing barge ST MARYS CEMENT (4679gt/1986) on passage Toledo, Ohio, to Bowmanville, Ontario. At dusk on leaving Lock No.4 of Welland Canal drifted astern due to wash from upbound vessel and struck arrester cable damaging mast and navigation equipment. Ship-arrester cable shear pins and linkage damaged.   2015: Still in service.

The **Uttern** (1515) on trials

(Cochrane archive)

Final preparations for the launch of the **Gösta** (1516).

(Ray Foster)

The **Frederic B Ingram** (1517) seen later in her career as the **Norman** of United Towing.

(Stuart Emery collection)

Rudderman (1519)

(Cochrane archive)

ROSS IMPLACABLE

The onlookers have dispersed and the majority of the shipyard workers have returned to more mundane tasks after the launch of *Ross Implacable* (1518).

(Roy Cressey collection)

The 1885-built tug **Cawood** assisted at the launch of **Ross Implacable**. Originally owned by Humber pilots, she was sold to John H Whitaker in 1912 and was converted from steam to diesel in 1953. When her commerical years were over, she was converted to a private yacht and sported two tall masts. Whitaker's tank barge **Jondor** is pulling alongside the trawler.

(Roy Cressey collection)

The **Ross Implacable** on trials.

(Cochrane archive)

The **Ross Implacable** in drydock.

(Author's collection)

DANGELD 337673 IMO 6826119 Motor refrigerated cargo ship	1521 22.11.1968 30.07.1969	694 355	225.7 94.7 22.5	Mirrlees Blackstone 2000bhp 2 x 8-cyl 13.75 knots	London & Rochester Trading Co Ltd, Rochester

8 passengers.   07.1969: Registered at Rochester.   1977: Sold to Mareantes Venturados Armadora S.A, Panama. Registered as a pallet carrier. Rochester registry closed. Registered at Panama as **REDSEA EXPRESS**.   1981: Sold to Ahmed Omer Al Majja Al Moudi Establishment, Jeddah, Saudi Arabia. Panama registry closed. Registered at Jeddah as **AL-MAJJA 1**.   1982: Sold to Star Navigation Co Ltd, Jeddah. Registered at Jeddah as **ALNEGMA ALKHADRA**.   1985: Sold to Purple Heart Shipping Co Ltd c/o Kalypso Atlantic Fishing, Athens. Jeddah registry closed. Registered at Limassol, Cyprus, as **MARITALIA**.   03.1992: Registered at Limassol as **GULF STAR**. 1998: Sold to East West Shipping c/o Nasser Ibrahimi Co, Dubai, United Arab Emirates (UAE).   1999: Sold to Hussein Mirza Mashaadi, La Paz, Bolivia. Limassol registry closed. Registered at La Paz as **ALLAH RASAN**.   2002: Sold to Najid M Shamsgharneh, Dubai, UAE. 09.2002: Registered at Wonsan, Peoples' Republic of Korea, as **MUBARAK ZAMEER**.   10.2005: Sold to Bassam Akram Ghazal, Deira, Dubai, UAE. Registered at Wonsan as **TABARK**.   12.2007: Sold to Ryadh M Gazal, Deira, Dubai, UAE. Wonsan registry closed. Registered at Freetown, Sierra Leone, as **NOOR AL RAHMAN**.   04.2010: Remeasured to 1427gt, 355net, 947dwt.   07.2011: Sold to undisclosed interests.   No further trace.

INVINCIBLE 339814 IMO 7016644 H96 Motor stern trawler freezer	1522 07.04.1970 25.09.1970	1085 482	215.0 39.6	English Electric 2160bhp 9-cyl 14.0 knots	Hudson Brothers (Trawlers) Ltd, Hull

Cost to build £1,222,000.   Government Grant £464,000.   Last of the four trawlers built to the design of the ROSS VANGUARD (see yard no.1503).   Ordered by Ross Group (deposit £73,450) and to have been named ROSS INVINCIBLE but completed for British United Trawlers Ltd (B.U.T.). Ross prefix dropped; first vessel to adopt B.U.T. livery.
25.09.1970: Registered at Hull (H96).   30.10.1970: Sailed on first trip to White Sea grounds.   1972: Top British freezer trawler landing 2692 tons.   11.1973: Transferred to Grimsby.   05.1978: B.U.T. transferred all their freezer trawler fleet to Hull.   09.02.1979: Sold to Wilemace Ltd, Hull.   08.10.1982: Sold to Seaboard Offshore (Diving & Pollution) Ltd, Tain, Ross-shire. Registered at Hull as **SEABOARD INVINCIBLE**. Converted to oil pollution recovery vessel.   17.04.1986: Hull registry closed. Registered at Inverness.   1988: Converted to offshore platform stand-by safety support vessel.   1991: Converted for survey support.   1996: Inverness registry closed. Registered at Hull. 1998: Sold to Hornbeck Offshore Ltd, Inverness. Registered at Hull as **HORNBECK INVINCIBLE**. Remeasured to 1461grt, 438net, 1727dwt.   05.1998: Sold to Tidewater Marine IOM Ltd, Douglas, Isle of Man.   11.1998: Hull registry closed. Registered at Port Vila, Vanuatu, as **SEAWAY INVINCIBLE**.   09.1999: Sold to BUE North Sea Ltd, Leith.   09.1999: Sold to Ceanic Worldwide Ltd, Panama. 2005: Sold to Hallstrom Holdings Pte. Ltd, Singapore. Port Vila registry closed. Registered at Panama.   01.2006: Sold to Eagle Shipping Ltd, Vaduz, Lichtenstein.   12.2007: Sold to Seaway Invincible Ltd, Singapore.   03.2014: Port Vila registry closed. Registered at Lomé, Togo.   2015: Still in service.

PACIFIC SHORE 337058 IMO 6900666 TM supply ship	1523 29.10.1968 28.02.1969	678 249	157.9 37.6 15.5	Blackstone 2400bhp 2 x 12-cyl 12.5 knots	Offshore Marine Ltd, Great Yarmouth

Built by Drypool Eng & Dry Dock Co at Hull.
12 passengers.   03.1969: Registered at London.   1980: Company and vessels sold to Zapata Gulf Marine Corp, Houston, Texas. Registered at London as **PACIFIC SERVICE**. Classed as tug supply ship.   1985: Sold to Caledonian Towage Co Ltd, Invergordon. London registry closed. Registered at Inverness as **KINNAIRD**.   1987: Converted to offshore platform standby safety vessel.   1988: Sold to George Craig & Sons Ltd, Aberdeen. Inverness registry closed. Registered at Aberdeen as **GRAMPIAN KESTREL**.   03.1991: Sold to Putford Enterprises Ltd, Lowestoft. Registered at Aberdeen as **PUTFORD TEAL**.   01.1996: Sold to Teal Shipping Ltd, St Peter Port, Guernsey. Aberdeen registry closed. Registered at Belize as **SEVEN HALELUYAH**.   02.2002: Sold to undisclosed owners.   08.2008: Sold to Moen Marine Shipping, Victoria Island, Lagos, Nigeria. Belize registry closed. Registered at Lagos as **SEA GIANT**.   2015: Believed still in service.

ARCTIC SHORE 337794 IMO 6906127 TM supply ship	1524 21.01.1969 28.05.1969	677 272	156.9 37.6 15.5	Lister Blackstone 2400bhp 2 x 12-cyl 12.5 knots	Offshore Marine Ltd, Rochester

12 Passengers.   05.1969: Registered at London.   1980: Company and vessels sold to Zapata Gulf Marine Corp, Houston, Texas. Registered at London as **ARCTIC SERVICE**. Classed as tug supply ship.   1985: Sold to Hutchfield Ltd, then Tradex Shipping Ltd, London. Registered at London as **RED SEA TRADER**.   1988: Sold to shipbreakers in Pakistan. Renamed **TOMOS TRADER** for delivery voyage, Assab - Gadani Beach.   17.02.1988: Arrived Gadani Beach and broken up by Sadaf Enterprises. Registry closed.

NOVA SHORE 337908 IMO 6914435 TM supply ship	1525 18.04.1969 08.08.1969	677 272	157.8 37.6 15.6	Mirrlees Blackstone 2400bhp 2 x 12-cyl 12.5 knots	Offshore Marine Ltd, Rochester

12 passengers.   08.1969: Registered at London.   1980: Company and vessels sold to Zapata Gulf Marine Corp, Houston, Texas. Registered at London as **NOVA SERVICE**. Classed as tug supply ship.   04.1988: Sold to Putford Enterprises Ltd, Britannia Marine PLC & Warbler Shipping Ltd, Lowestoft. Converted to offshore platform standby safety vessel. Registered at London as **NOVA**.   2002: Sold to Viking Standby, Montrose & Putford Enterprises Ltd, Lowestoft.   2004: Sold to Boston Putford Offshore Safety Ltd & BUE Viking Ltd c/o Seacor SNS, Lowestoft.   24.11.2011: Sold to shipbreakers and broken up. Registry closed.

STIRLINGBROOK 338918 IMO 6927561 Motor cargo ship	1526 12.09.1969 12.1969	1597 1139 2947	265.0 42.0 20.5	English Electric (Ruston) 2100bhp 8-cyl 12.0 knots	County Ships Ltd, London

Ice strengthened.   12.1969: Registered at London.   1983: Sold to Medimar S.p.A, Castellammare di Stabia, Italy. London registry closed. Registered at Castellammare di Stabia as **RECOMONE**.   1984: Sold to Monremar di Navigazione SAS, Naples. Remeasured to 901net. 24.06.1987: Vessel badly damaged and disabled by engine room fire, whilst on passage from Sfax towards Ravenna.   30.06.1987: Arrived Malta under tow.   07.1987: Sold to Acciaierie e Ferriere di Porto Nogaro SpA, Porto Nogaro, for breaking up.   23.07.1987: Arrived Porto Nogaro and broken up. Registry closed.

Dangeld (1521)

(Cochrane archive)

The **Pacific Shore** (1523) at Great Yarmouth on 17 April 1976.

(Bernard McCall)

Arctic Shore (1524)

(Cochrane archive)

SUSSEXBROOK	1527	1596	265.0	English Electric (Ruston)	County Ships Ltd,
339073 IMO 7005140	11.12.1969	1131	42.0	2100bhp 8-cyl	London
Motor cargo ship	15.04.1970	2947	20.5	12.0 knots	

Ice strengthened.   04.1970: Registered at London.   1983: Sold to Medimar S.p.A, Castellammare di Stabia, Italy.   London registry closed. Registered at Castellammare di Stabia as **IERANTO**.   1987: Sold to Gestione Navi S.r.l, (GE.WA), Castellammare di Stabia.   Registered at Castellammare di Stabia as **IDA ERRE**.   1990: Sold to Sonar S.r.l, Piano di Sorrento, Italy.   Registered at Castellammare di Stabia as **SONAR PRIMA**.   1995: Sold to Silver Mast Shipping Ltd, Valletta, Malta.   Castellammare di Stabia registry closed.   Registered at Kingstown, St Vincent & the Grenadines as **GOLD STAR 1**.   Remeasured to 1778gt 1108net.   11.1996: Sold to Bulk Traders International, Beirut, Lebanon.   Registered at Kingstown as **EL CONDOR**.   1997: Registered at Kingstown as **MEGANE**.   1998: Sold to Transindo Shipping Ltd.   Registered at Kingstown as **FROSINA**.   04.1998: Sold to Med Express Inc, Kingstown.   06.1998: Sold to Kneo Perla, Durres, Albania.   Kingstown registry closed.   Registered at Durres.   01.2003: Sold to Zotaj Shipping & Trading Co Ltd, Durres, Albania. 07.2005: Sold to Gjon Shtjefni, Durres, Albania.   Registered at Durres as **SHKODRA**.   12.2006: Durres registry closed.   Registered at Freetown, Sierra Leone.   04.2007: Freetown registry closed.   Registered at Durres.   08.2008: Sold to Turkish owners, not identified. 04.08.2008: Sold to Letfallah Shipping Co, Tartous, Syria.   Durres registry closed.   08.2008: Registered at Wonsan, Peoples' Republic of Korea, as **MUSTAFA**.   12.2009: Wonsan registry closed.   Registered at Zanzibar, Tanzania.   05.2010: Sold to Gemi Yan Sanayi, Istanbul. 27.05.2010: Beached at Aliaga and broken up.   Registry closed.

ELOQUENCE	1528	392	139.0	Bergius Kelvin	Lomdon & Rochester Trading
337676 IMO 6920006	03.06.1969	278	25.5	320bhp 8-cyl	Co Ltd,
Motor cargo ship	07.1969	591	12.4	9.0 knots	Rochester

07.1969: Registered at Rochester.   1980: Re-registered to London & Rochester Trading Co Ltd (Crescent Shipping), Rochester. 1985: Sold to Brian Thomas Cuckow, Gillingham.   Registered at Rochester as **GORE**.   03.1987: Sold to Dennison Shipping Ltd, Kirkwall. Rochester registry closed.   Registered at Kirkwall as **HOLM SOUND**.   1995: Sold to Gardscreen Shipping Ltd, Rainham, Gillingham. 1996: Sold to Radmoor Shipping Ltd, Wolverhampton. Limited load-line.   1997: Laid up at Otterham Quay.   10.1997: Sold to J J Prior (Transport) Ltd, Fingringhoe, Essex. Bought and refurbished at Acorn Shipyard, Rochester, with the aid of a Freight Facilities Grant. To operate Fingringhoe-Thames with sand.   04.1998: Kirkwall registry closed.   28.04.1998: Registered at Colchester as **PETER PRIOR**. 2012: Stripped and used as a hulk.   26.07.2013: Broken up at Greenhithe.   Registry closed.

ISLAND SHORE	1529	694	160.4	Lister Blackstone Mirrlees	Offshore Marine Ltd,
339014 IMO 7005322	01.12.1969	232	37.5	4000bhp 2 x 16-cyl	London
TM supply ship	06.03.1970	744	15.5	13.5 knots	

Built by Drypool Eng & Dry Dock Co Ltd, Hull.   12 passengers. 03.1970: Registered at London.   1977: Converted to a tug supply ship - 720grt, 244net.   1980: Company and vessels sold to Zapata Gulf Marine Corp, Houston, Texas.   Registered at London as **ISLAND SERVICE**.   1987: Sold to Lethbridge Maritime Co S.A c/o Ocean Offshore Pty, Cape Town, South Africa.   London registry closed.   Registered at Panama.   Remeasured to 717gt, 232net.   1997: Sold to Dive & Survey Vessels Ltd c/o Smit Octo Marine Ltd, Cape Town.   Panama registry closed.   Registered at Kingstown, St Vincent & the Grenadines. 08.1999: Sold to O.I.S. International, Port Harcourt, Nigeria.   06.2006: Kingstown registry closed.   Registered at Port Harcourt as **RIVERS SUCCESS**.   2014: No further information.

CAPE SHORE	1530	694	160.2	Lister Blackstone Mirrlees	Offshore Marine Ltd,
339113 IMO 7005229	16.12.1969	232	37.6	4000bhp 2 x 16-cyl	London
TM supply ship	04.05.1970	744	15.5	13.5 knots	

Sub-contracted and built by Chas Hill & Sons Ltd, Bristol (yard no.462). 12 passengers.   05.1970: Registered at London.   1979: Converted to a tug supply ship - 770grt, 262net.   1980: Sold to Texas Instruments Ltd, London.   Registered at London as **R W OLSON**. Fitted out for research work - 702grt, 210net.   1988: Sold to Intership Ltd, U.K. Registered at London as **MARI HOLM**.   1988: Sold to Svitzer Ltd, Gt Yarmouth.   Registered at London as **SVITZER MERCATOR**. 02.1994: Sold to Sea Shepherd Conservation Society.   London registry closed.   Classed as a motor yacht and registered at Belize as **WHALES FOREVER**.   07.1994: Extensive bow damage sustained in an incident involving Norwegian Coastguard vessel K/V ANDANES in Norwegian territorial waters.   1995: Sold to Saint Somewhere Island Tours, Hawaii.   Registered at Belize as **ST SOMEWHERE I**. 1998: Sold to Caldwell Diving Inc, Hawaii.   Registered at Belize as **THOMAS C**.   1999: Belize registry closed.   Registered at San Lorenzo, Honduras. Converted to a cable layer.   2000: Sold to International Telecom, Pointe-Claire, Quebec, Canada.   San Lorenzo registry closed. Registered at Nassau, Bahamas.   01.2001: Sold to Cable Ventures Inc, San Antonio, Texas, USA.   Nassau registry closed.   Registered at Belize.   2003: Sold to Caldwell Marine International Inc, New Jersey, USA.   Belize registry closed.   Registered at Nassau.   2004: Sold to Marika Investments Ltd, San Antonio, Texas, USA.   Nassau registry closed.   Registered at San Lorenzo as **FENDER CARE 2**. 07.2006: Sold to Asteria Navigation Inc, Monrovia, Liberia.   2011: San Lorenzo registry closed.   Registered at Basseterre, St Kitts & Nevis. 2013: Renamed **AFRICA SUPPORT 2**.   No port of registry recorded.   No further information.

IMPERIAL SERVICE	1531	691	164.0	English Electric	Zapata Offshore Services
341236 IMO 7027265	25.07.1970	262	38.5	7040bhp 2 x 16-cyl	Ltd,
TM tug supply ship	03.02.1971	950	17.0	16.0 knots	London

Built by Drypool Engineering & Dry Dock Co Ltd, Hull. 02.1971: Registered at London.   1980: Transferred to Zapata Marine Services Inc, Panama.   London registry closed.   Registered at Panama.   1985: Sold to Adel Mahmoud Mohamed, Egypt.   Registered at Panama as **IMPERIAL**, then **IMPERIAL SERVICE**, then **ABU EL HOOL 7**.   1986: Sold to Pyramids Maritime Services Co, Egypt.   1986: Sold to Pharaonic Petroleum Services Co, Suez, Egypt. 11.1988: Sold to Indian shipbreakers and broken up at Gadani Beach.   12.1988: Registry closed.

PARAMOUNT SERVICE	1532	692	164.0	English Electric	Zapata Offshore Services
342873 IMO 7030406	15.09.1970	262	38.5	7040bhp 2 x 16-cyl	Ltd,
TM tug supply ship	05.1971	834	17.0	16.0 knots	London

Sub-contracted and built by Charles Hill & Sons Ltd, Bristol, (yard no.463). 05.1971: Registered at London.   1981: Sold to Baani Shipping Co Ltd, Bombay, India.   London registry closed.   Registered at Bombay as **SRI MAHAVIR**.   16.04.1981: Returned to Barcelona after developing engine trouble during her delivery voyage to Bombay. Laid up. 01.1986: Sold by auction to Barcelona ship repairers Talleres Nuevo Vulcano.   1988: Sold to shipbreakers and broken up. Registry closed.

STEERSMAN 341182 IMO 7026431 Motor oil products tanker	1533 21.07.1970 15.12.1970	1567 953 2932	260.0 41.0 20.0	Brons 1860bhp 16-cyl 12.0 knots	Helmsman Shipping Co Ltd, London

12.1970: Registered at London. 1973: Owners restyled C Rowbotham & Sons (Management) Ltd, London. 1976: Registered at London as **RIVER SHANNON**. 1981: Sold to Rowbotham Tankships Ltd, London. 1988: London registry closed. Registered at Douglas, Isle of Man. 1992: Sold to Rowbotham Tankships (Gibraltar) Ltd, Gibraltar. 01.1993: Sold to Skopelos Shipping Co Ltd, Piræus, Greece. 02.1993: Douglas registry closed. Registered at Piræus as **SKOPELOS**. 03.1997: Sold to Al Faihaa Trading L.L.C, Dubai, United Arab Emirates. 04.1997: Piræus registry closed. Registered at Mina Khalid as **AL BATOOL**. Remeasured to 1672gt, 1018net, 2979dwt. 1998: Sold to Kamel H Mohamed. No port of registry recorded. 2003: Sold to Indian shipbreakers. 07.05.2003: Beached at Alang and broken up. Registry closed.

MONARCH SERVICE 342984 IMO 7110050 TM tug supply ship	1534 28.10.1971 01.1972	691 262 859	163.0 38.5 17.0	Ruston Paxman 3520bhp 2 x 16-cyl 16.0 knots	Zapata Offshore Services Ltd, Great Yarmouth

Built by Drypool Engineering & Dry Dock Co Ltd, Hull.
8 passengers. 01.1972: Registered at London. 17.02.1977: Foundered in North Sea after taking on a list and turning turtle while taking equipment off drilling platform BORGNY DOLPHIN in position 58.12N 0.07E. All crew saved.

MAJESTIC SERVICE 358480 IMO 7206005 TM tug supply ship	1535 1972 06.1972	688 261 844	163.0 38.6 17.0	English Electric 7040bhp 2 x 16-cyl 16.5 knots	Zapata Offshore Services Ltd, Great Yarmouth

Sub-contracted and built by Charles Hill & Sons Ltd, Bristol (yard no.469).
06.1971: Registered at London. 1985: Transferred to Zapata Marine Services S.A, Panama. London registry closed. Registered at Panama. 1986: Transferred to Offshore Marine Ltd, Great Yarmouth. 1988: Sold to shipbreakers and broken up. Registry closed.

SOMERSETBROOK 341268 IMO 7037131 Motor cargo ship	1536 14.11.1970 03.1971	1596 1131 2947	265.0 42.0 20.5	Ruston Paxman 2100bhp 8-cyl 12.0 knots	County Ships Ltd, London

Ice strengthened. 03.1971: Registered at London. 1983: Sold to Fabula Shipping Ltd, Piræus. London registry closed. Registered at Limassol, Cyprus, as **FIRMUS**. Remeasured to 1588gt, 1148net. 1985: Sold to Societa di Navigazione Vegemar, Naples, Italy. Limassol registry closed. Registered at Naples as **SUNRAY**. 1987: Sold to Albamar S.a.S di Luigi Gargiulo & Co, Naples. 1988: Sold to Franco de Paolis S.p.A, Naples. 1988: Registered at Naples as **DEPASPED**. 07.1889: In port damaged - to be broken up. 05.1990: Sold to Fin Par S.p.A, Naples. Repaired. 10.1992: Sold to Terranova Maritime S.A, Panama. Naples registry closed. Registered at Panama as **SPED**. Remeasured to 1778gt, 1108net, 2947dwt. 06.1994: Sold to Vassin Kassab & Zuheir Joud, Lattakia, Syria. Panama registry closed. Registered at Lattakia as **MANDARIN**. 1997: Sold to Zuhair Dib Joud, Naima Ahmad Ismail & Mohammad Abdullah Baizid, Lattakia. 2000: Sold Veral S.A, Turkey. 09.10.2000: Arrived and beached at Aliaga, Turkey, for breaking up. Registry closed.

SURREYBROOK 342850 IMO 7108150 Motor cargo ship	1537 11.05.1971 09.1971	1596 1131 2947	265.0 42.0 20.5	English Electric 2100bhp 8-cyl 12.75 knots	County Ships Ltd, London

Ice strengthened. 09.1971: Registered at London. 1982: Sold to Naviera Romana S.A. c/o Lineas Maritimas de Santo Domingo SA, Santo Domingo, Dominican Republic. London registry closed. Registered at Santo Domingo as **ROMANA**. 09.1990: Sold to Next Generation Shipping Co Ltd c/o Anteus Shipping Co Ltd, Piræus. Santo Domingo registry closed. Registered at Valletta, Malta, as **LITO**. 03.1991: Sold to Sea Wing Shipping Ltd, Piræus. Registered at Valletta as **GOD SPIRIT**. 06.1993: Sold to Artline Maritime Ltd, Piræus. Registered at Valletta as **AQUARIUS**. 06.1994: Sold to Serenade Enterprises Ltd, Piræus. Valletta registry closed. Registered at Kingstown, St. Vincent & the Grenadines, as **SERENADE**. 1998: Registered at Kingstown as **CITY OF LONDON**. 2003: Sold to Demka Ltd, Turkey, for breaking up. 12.09.2003: Arrived Aliaga. Registry closed.

BAY SHORE 342942 IMO 7117175 TM tug supply ship	1538 23.07.1971 07.11.1971	694 232 766	160.2 37.5 15.5	Ruston Paxman 5280bhp 2 x 12-cyl 13.25 knots	Offshore Marine Ltd, London

Contracted to and built by J Bolson & Son Ltd, Poole (yard no.572).
Ice strengthened. 12 passengers. 11.1971: Registered at London. 1977: Sold to Dome Petroleum Ltd c/o Canadian Marine Drilling Ltd, Calgary, Alberta, Canada. London registry closed. Registered at Vancouver as **CANMAR SUPPLIER VI**. 1991: Sold to Amoco Canada Resources Ltd c/o Canadian Marine Drilling Ltd, Calgary, Alberta, Canada. 02.1994: Sold to Intership Ltd c/o Papaipannou Brothers Shipping Enterprise Co Ltd, Piræus, Greece. Vancouver registry closed. Registered at Kingstown, St. Vincent & the Grenadines, as **SUPPLIER VI**. 01.1995: Sold to Iran Marine Services Co, Tehran Iran. Kingston registry closed. Registered at Dammam, United Arab Emirates, as **YAZD**. 11.1997: Sold to Pars Dolphin of Qeshm, Bandar Bushehr, Iran. 06.2014: Sold to United Captain Group, Alexandria, Egypt. Registered at Dar es Salaam, Tanzania, as **LMAR**. 2015: Believed still in service.

POLAR SHORE 343013 IMO 7101619 TM tug supply ship	1539 27.02.1971 20.12.1971	700 293 778	50,81 11,76 5,12	Ruston Paxman 7083bhp 2 x 12-cyl CPP 12.0 knots	Offshore Marine Ltd, London

Built by Drypool, Selby. First Cochrane vessel designed and built to metric dimensions. All vessels metric from this yard number.
Ice strengthened. 12 passengers. 12.1971: Registered at London. 1977: Sold to Dome Petroleum Ltd c/o Canadian Marine Drilling Ltd, Calgary, Alberta, Canada. London registry closed. Registered at Vancouver as **CANMAR SUPPLIER VII**. 1991: Sold to Amoco Canada Resources Ltd c/o Canadian Marine Drilling Ltd, Calgary, Alberta, Canada. 02.1995: Sold to Ocean Navigation Inc, Quebec c/o Les Remorqueurs du Quebec Ltde, Quebec. Vancouver registry closed. Registered at Quebec as **OCEAN FOXTROT**. 2001: Sold to Groupe Ocean Inc, Quebec. 2008: Sold to Ocean Remorquage Trois, Trois-Rivières, Quebec, c/o Ocean Group Inc, Quebec. 10.2004: Sold to undisclosed interests. 10.2014: No flag recorded. 2015: Believed still in service.

Sussexbrook *(1527)*

(Cochrane archive)

Steersman *(1533)*

(Cochrane archive)

Monarch Service *(1534)*

(Cochrane archive)

HELMSMAN 343160 IMO 7123370 Motor oil tanker	1540 05.11.1971 15.04:1972	3705 2308 6264	97,54 14,94 7,93	Ruston Paxman 3520bhp 16-cyl CPP 13.0 knots	Helmsman Shipping Co Ltd, London	

Built by Drypool, Selby.
04.1972: Registered at London. 1972: Sold to C Rowbotham & Sons (Management) Ltd, London. 1981: Sold to Rowbotham Tankships Ltd, London. 1987: London registry closed. Registered at Douglas, Isle of Man. 1992: Sold to Rowbotham Tankships (Gibraltar) Ltd, Gibraltar. Douglas registry closed. Registered at Gibraltar. 01.02.1993: Sold to P&O Tankships (Gibraltar) Ltd, Gibraltar. 03.1994: Sold to Galana Petroleum Ltd, Mombasa, Kenya. Registered at Gibraltar as **RUFIJI**. 05.1997: Sold to Seagull Maritime Ltd, Mombasa. 06.1997: Gibraltar registry closed. Registered at Nassau, Bahamas as **SEAGULL**. 04. 2003: Sold to Indian shipbreakers and broken up at Mumbai. Registry closed.

SOLENTBROOK 358547 IMO 7214882 Motor cargo ship	1541 15.05.1972 27.07.1972	1597 1131 2982	80,78 12,81 6,25	Ruston Paxman 2100bhp 8-cyl 12.0 knots	County Ships Ltd, London	

Built by Drypool, Selby.
Ice strengthened. 07.1972: Registered at London. 1982: Sold to Kemp Navigation Co Ltd c/o Liakos Maritime Co Ltd, Piræus. London registry closed. Registered at Limassol as **STAVROS H**. 1994: Sold to Kanoria Shipping Ltd, Piræus, Greece. Limassol registry closed. Registered at Kingstown, St. Vincent & the Grenadines, as **TAVROS II**. 09.1995: Sold to Marissa Shipping Ltd, Monrovia, Liberia c/o Trinto Maritime Co, Perama, Greece. Registered at Kingstown as **ELPIDA**. 05.1996: Registered at Kingstown as **LENTBRO**. 1998: Registered at Kingstown as **TRITON**. 1999: Sold to Jetty Shipping Inc. 01.2001: Sold to Unimed International, Athens, Greece. Registered at Kingstown as **THEOFILOS**. 07.2003: Sold to Beam Shipping & Trading Ltd c/o Ismail Shipping, Constantza, Romania. Kingstown registry closed. Registered at Belize as **LAMIAA**. 03.2004: Sold to Ahmed Mohamad Libadi c/o Labbadi Ship Management, Banias, Syria. 05.2004: Registered at Belize as **RASHA MOON**. 08. 2004: Registered at Belize as **HAJE KHEREH**. 13.08.2007: Sold to Al Rahal Shipping Co Ltd c/o Majid Abdulla Shipping LLC, Dubai, United Arab Emirates. Belize registry closed. 08.2008: Registered at Kingstown, St Vincent & the Grenadines, as **AI RAHALAH 1**. 12.2010: No flag recorded. 2015: Believed still in service.

NELLIE M 3431377 IMO 7204409 Motor cargo ship	1542 02.02.1972 14.04.1972	783 448 1180	57,61 10,21 4,57	W H Allen 1160bhp 8-cyl 12.0 knots	Metcalf Motor Coasters Ltd, London	

Built by Drypool, Selby.
04.1972: Registered at London. 1972: Company and ships sold to Booker Line Ltd, Liverpool. 1978: Sold to Coe Metcalf Shipping Ltd, Liverpool. 1978: Lengthened by 36ft to 225.0 ft - 954grt, 594net, 1393dwt. 06.02.1981: At anchor off Moville, Lough Foyle, awaiting favourable weather to proceed to Coleraine, cargo 1260tons of coal loaded at Blyth. Boarded by IRA, explosive charges placed in engine room. Crew ordered into lifeboat and towed clear of the ship and cast off near shore at which time two explosions were heard and flames engulfed the after end. Shortly afterwards the ship started to settle by the stern and founder. 12.07.1981: After delay in appointing salvors, refloated by Eurosalve Ltd, Folkestone, towed to Londonderry for discharge of cargo. Declared a CTL; abandoned by underwriters to salvors. 03.1982: Sold in damaged condition to Irish principals for breaking up. Decided to repair and trade. 08.1982: After repair at Londonderry registered to Lofoten Cia de Nav Panama as **ELLIE**. Sailed for Dublin and drydocking. 06.03.1983: Laid up at Dundalk. 26.04.1983: Sailed for Belfast. 08.1983: Arrested in Spain but sailed without assistance. 1984: Sold to Phoenix Offshore Ltd, Wadebridge, Cornwall. 1984: Sold to Timrix Shipping Co Ltd, Hull, with delivery at Hull. 03.1985: Panama registry closed. Registered at Hull as **TIMRIX**. 1991: Hull registry closed. Registered at Nassau, Bahamas. 1994: No.2 hatch coaming raised, remeasured to 1007gt, 617net, 1393dwt. 11.1995: Sold to Apex Maritime Ltd c/o Seaflight Management Consultants Ltd, Dulwich, London. Hull registry closed. Registered at Valletta, Malta, as **MALTESE VENTURE**. 11.1996: Sold to Modern Marine Operations Ltd, Jersey, Channel Islands, c/o Pro Chart BV, Rotterdam. Registered at Valletta as **SPEZI**. 1997: Sold to Simon J Lyon-Smith, Crediton, & others. 1997: Sold to Maritima Santa Catalina, Isla de San Andres, Columbia. 07.1998: Sold to Caribbean Island Shipping Inc, Belize. Valletta registry closed. Registered at Belize as **DOVE**. 11. 2000: Sold to St. Marten Ltda, Cartagena de Indias, Colombia. Registered at Belize as **AMAZON'S DOLPHIN**. 05.2009: Reported sold to undisclosed interests. Belize registry closed. Registered at Kingstown, St Vincent & the Grenadines, as **OCEANIC LADY**. 03.2011: Kingstown registry closed. Registered at São Tomé, São Tomé & Principe, as **CARMEN II**. 2015: Believed still in service.

ASTRAMAN 358853 IMO 7229306 Motor chemical tanker	1543 26.10.1972 06.03.1973	1599 1079 3202	82,11 13,71 6,10	English Electric Ruston Paxman 3520bhp 2 x 8-cyl CPP 14.0 knots	Helmsman Shipping Co Ltd, London	

Built by Drypool, Hull.
03.1973: Registered at London. 1973: Sold to C Rowbotham & Sons (Management) Ltd, London. 1981: Sold to Rowbotham Tankships Ltd, London. 1987: London registry closed. Registered at Douglas, Isle of Man. 1993: Sold to Rowbotham Tankships (Hong Kong) Ltd, Hong Kong. Douglas registry closed. Registered at Hong Kong. 01.02.1993: Sold to P&O Tankships (Hong Kong) Ltd, Hong Kong. 07.1995: Sold to Pacific Dynamic Ltd, Hong Kong. Registered at Hong Kong as **STRAMAN**. 11.1995: Sold to Ihlas Finans Kurumu A.S, Istanbul, Turkey. Hong Kong registry closed. Registered at Istanbul as **FERMAN I**. 12.1999: Sold to Euro Oil Ltd, Bolivia. Renamed **ARABIAN QUEEN** (no port of registry recorded). 11.2000: Sold to Gulf Oil Co LLC, Ajman, United Arab Emirates. Registered at Batumi, Georgia. 2001: Sold to Indian shipbreakers. 12.02.2001: Arrived Mumbai for breaking up. Registry closed.

POLARISMAN 360752 IMO 7304390 Motor chemical tanker	1544 19.01.1973 08.1973	1599 1079 3151	82,11 13,71 6,10	English Electric Ruston Paxman 3520bhp 2 x 8-cyl CPP 14.0 knots	C Rowbotham & Sons (Management) Ltd, London	

Built by Drypool Engineering & Dry Dock Co Ltd, Hull.
08.1973: Registered at London. 1981: Sold to Rowbotham Tankships Ltd, London. 1987: London registry closed. Registered at Douglas, Isle of Man. 1992: Sold to Rowbotham Tankships (Gibraltar) Ltd, Gibraltar. Douglas registry closed. Registered at Gibraltar. 01.02.1993: Sold to P&O Tankships (Gibraltar) Ltd, Gibraltar. 07.1994: Sold to Milbank Ltd, part of Jaisu Shipping Co Pvt Ltd, Kandla, India. Gibraltar registry closed. Registered at Kingstown, St Vincent & the Grenadines as **GOD PRESTIGE**. 09.09.2011: Continued existence in doubt.

WATERCOURSE 903026 IMO 7340796 Y.30 Motor water carrier	1545 03.05.1973 06.12.1973	285 86 300	37,49 7,56 3,51	Mirrlees Blackstone 600bhp 8-cyl 11.0 knots	Ministry of Defence (Navy), Royal Maritime Auxiliary Service, London

Built by Drypool Engineering & Dry Dock Co Ltd, Hull.
1988: Sold to Rosyth Marine Services, Dunfermline. Registered at Leith as **RMS WATERCOURSE**. 01.05.2012: Sold to Michael Pratt, Rainham, Kent. 2015: Still in service.

WATERFOWL IMO 8943870 Y.31 Motor water carrier	1546 29.08.1973 27.03.1974	288 86 300	37,49 7,56 3,51	Mirrlees Blackstone 600bhp 8-cyl 11.0 knots	Ministry of Defence (Navy), Royal Maritime Auxiliary Service, London

Built by Drypool Engineering & Dry Dock Co Ltd, Hull.
05.1999: Sold to Coloured Fin Ltd, Chaguaramas, Trinidad and Tobago. Registered at Port of Spain as **THISTLE**. 2015: Still in service.

TODRA IMO 7340801 Motor tanker	1547 12.12.1973 27.06.1974	1599 698 2444	79,26 12,43 5,29	Motorenwerke 2400bhp 8-cyl 13.0 knots	Société Shell du Maroc & Mobil Oil Maroc, Mohammedia, Morocco

Built by Cochrane & Sons Ltd, Selby (Drypool Group Ltd).
07.1974: Registered at Mohammedia. 1986: Sold to Société de Cabotage Petrolier, Mohammedia. 11.1995: Sold to Tropicana Shipping Co Ltd, Valetta, Malta. 1995: Remeasured to 1807gt, 693net, 2406dwt. 1995: Mohammedia registry closed. Registered at Valletta as **SUPER TRADER**. 01.09.1996: Sold to Tramaco S.A. (Transportadora Maritima de Combustible), Montevideo, Uruguay. 1996: Valletta registry closed. Registered at Montevideo as **SUPER T**. 2015: Still in service.

SEAFORTH CHALLENGER 359078 IMO 7101619 TM tug supply ship	1548 03.05.1973 10.1973	733 282 1040	49,01 11,83 5,21	Mirrlees Blackstone 5000bhp 2 x 16-cyl CPP 14.0 knots	Seaforth Maritime Ltd, Aberdeen

Built by Cochrane & Sons Ltd, Selby (Drypool Group Ltd).
12.10.1973: Registered at Aberdeen. 1977: Sold to Risdon Beazley Marine Ltd, Southampton. Aberdeen registry closed. Registered at Southampton as **SEAFORD**. 1981: On closure of company sold to Smit International South East Asia (Pte) Ltd, Singapore. Southampton registry closed. Registered at Nassau, Bahamas, as **SMIT MANILA**. Classed as a diving support vessel - 791gt, 258 net. 1989: Sold to Swedish Marinen. Converted by Pan United Shipyards, Singapore, to a depot ship. On completion commissioned as HSwMS **UTÖ**.
08. 2003: Sold to Zakher Marine International Inc, Abu Dhabi, United Arab Emirates. Registered at Kingstown, St. Vincent & the Grenadines as **ZAKHER MOON**. 2014: Remeasured to 1203gt. 04. 2006: Sold to Maridrive Oil & Services (SAE), Giza, Egypt. Kingstown registry closed. Registered at Batumi, Georgia, as **MOP 50**. 06.2007: Sold to Meridive Offshore Projects, Alexandria c/o Maridive Oil & Services (SAE), Giza. 06.2013: Sold to Star Global Energy FZE c/o Star Petroleum Co FZC, Ajman, United Arab Emirates. Batumi registry closed. Registered at Malakal as **STAR GLOBAL**. 2015: Still in service.

SEAFORTH PRINCE 350975 IMO 7319785 TM tug supply ship	1549 17.05.1973 08.1973	732 282 1040	49,01 11,83 5,21	Mirrlees Blackstone 5000bhp 2 x 16-cyl CPP 14.0 knots	Seaforth Maritime Ltd, Aberdeen

Contracted to and built by Chas Hill & Sons Ltd, Bristol (yard no.473).
20.08.1973: Registered at Aberdeen. 1976: Sold to Lloyds & Scottish Development Ltd, Edinburgh (Seaforth Maritime Ltd). 1981: Leased to Zapata Offshore Services Ltd, Gt Yarmouth. Aberdeen registry closed. Registered at London as **CHAPARRAL SERVICE**.
1983: Sold to Zapata Offshore Services Ltd, Gt Yarmouth. 1988: Sold to Thai Hua Lee Co Ltd, Thailand, for breaking up.
11.1988: Arrived Samut Prakan and broken up. Registry closed.

SEAFORTH HERO 359073 IMO 7314876 TM tug supply ship	1550 23.01.1973 03.07.1973	733 282 1040	49,01 11,83 5,21	Mirrlees Blackstone 5000bhp 2 x 16-cyl CPP 14.0 knots	Seaforth Maritime Ltd, Aberdeen

Built by Drypool Eng & Dry Dock Co Ltd, Hull.
05.07.1973: Registered at Aberdeen. 1976: Sold to Lloyds & Scottish Development Ltd, Edinburgh. 1987: Sold to Barnacle Shipping Co Ltd c/o Dayak Corp, Piræus. Registered at Aberdeen as **STRAIT II**. 1990: Sold to Istanbul Gemi Kurtarma ve Insaat Enkaz Kaldirma Sanayi ve Ticaret Ltd, Sirketi (Salvage, Construction and Wreck Removal Ltd), Istanbul, Turkey. Aberdeen registry closed. Registered at Istanbul as **OCEAN ASLI** (also classed as an anchor handling vessel). 04.1992: Sold to Erol Sonmezler, Istanbul. Registered at Istanbul as **BALINA I**. 02.1993: Sold to Istanbul Gemi Kurtarma ve Insaat Enkaz Kaldirma Sanayi ve Ticaret Ltd, Sirketi (Salvage, Construction and Wreck Removal Ltd), Istanbul, Turkey. 07.1995: Sold to S H Supply S.A, Suez, Egypt. Istanbul registry closed. Registered at Panama as **RAMSIS THE SECOND**. Remeasured to 1176gt, 705net, 1040dwt. 01.1998: Sold to Zakher Marine International Inc, Abu Dhabi, United Arab Emirates. 10.1999: Panama registry closed. Registered at Kingstown, St. Vincent & the Grenadines, as **ZAKHER XII**.
08.04.2008: Sold to Consolidated Discounts Ltd, Calabar, Nigeria. Registered at Kingstown as **POLARIS EHCO**. 15.07.2009: Laid up. No flag declared.

SEAFORTH CHIEFTAIN 359982 IMO 7330325 TM tug supply ship	1551 13.09.1973 09.02.1974	777 302	52,00 11,80 5,21	Mirrlees Blackstone 5000bhp 2 x 16-cyl CPP 14.0 knots	Seaforth Maritime Ltd, Aberdeen

Built by Cochrane & Sons Ltd, Selby (Drypool Group Ltd).
05.02.1974: Registered at Aberdeen. 1975: Sold to Royal Bank Leasing Co, Edinburgh (Seaforth Maritime Ltd). 1976: Sold to R B Leasing Co, Edinburgh (Seaforth Maritime Ltd). 1982: Sold to Northern Offshore Logistica (N.O.L.), Genova, Italy. Aberdeen registry closed. Registered at Genova as **ANTARES**. 1985: Sold to N.O.L. Italia S.p.A. & Societa Investimenti Amatorali, Genova. 1986: Genova registry closed. Registered at Ancona. 1988: Ancona registry closed. Registered in Siracusa. 1992: Sold to Offshore Adriatica S.r.L, Siracusa. 2004: Sold to Med Offshore SpA, Naples, Italy. Siracusa registry closed. Registered at Naples as **MED CINQUE**.
09.2005: Sold to Star Marine Services SA c/o Asetanian Marine Pte, Ltd, Singapore. Naples registry closed. 09.2005: Registered at Panama as **ATLANTIC CHALLENGER**. 08.2011: Panama registry closed. Registered at Jakarta. 10.2011: Remeasured to 918gt, 302net, 1112dwt. 09.11.2011: Sold to Bayu Maritim Berkah Pte, Jakarta, Indonesia. 2015: Still in service.

Polarisman*(1544)*

(Cochrane archive)

The **Waterfowl** *(1546) enters the lock at Hull on completion of trials.*

(Cochrane archive)

Seaforth Prince *(1549)*

(Cochrane archive)

MARIS 1	1035 (Bev)	358	33,51	Mirrlees	Government of Nigeria Federal
321915 IMO 7329493	31.08.1973	75	9,00	2200bhp 2 x 8-cyl	Ministry of Transport -
TM patrol vessel	29.03.1974	72	3,51	12.0 knots	Maritime Division,
					Lagos, Nigeria

Built by Charles D Holmes & Co Ltd, Beverley (yard no.1035). Patrol vessel, West African coasting service, Cape Verde to the River Congo, not more than 50 miles offshore.
06.1974: Registered at Lagos. 2015: Still in service.

STIRLING ROCK	1552	699	51,01	W H Allen	Stirling Shipping Co Ltd,
361646 IMO 7340813	25.03.1974	340	11,64	2500bhp 2 x 8-cyl	Glasgow
TM supply ship	30.10.1974	907	4,65	12.5 knots	

Launched at Beverley; completed by Cochrane & Sons Ltd, Selby (Drypool Group Ltd).
11.1974: Registered at Glasgow. 1978: Sold to R.B. Leasing Co, Edinburgh (Stirling Shipping Co Ltd). 1984: Sold to Royal Bank Leasing Co, Edinburgh. Registered for 12 passengers. 1984: Sold to Haven Shipping Co Ltd, Glasgow. Registered at Glasgow as STONEHAVEN. Converted to offshore platform standby safety vessel - 783gt, 377net. 1988: Sold to George Craig & Sons, Aberdeen. Glasgow registry closed. Registered at Aberdeen as GRAMPIAN SHIELD. 1995: Remeasured to 789gt, 377net, 907dwt. 06.2004: Sold to Caribbean Supplies Panama Corp, Panama City, Panama. 06.2004: Aberdeen registry closed. Registered at Panama as LITTLE JOE.
06.2011: Registry closed - "Broken up".

STIRLING BRIG	1553	699	51,01	W H Allen	Stirling Shipping Co Ltd,
361642 IMO 7342237	26.03.1974	299	11,61	2500bhp 2 x 8-cyl	Glasgow
TM supply ship	03.09.1974	923	4,65	12.5 knots	

Built by Cochrane & Sons Ltd, Selby (Drypool Group Ltd).
09.1974: Registered at Glasgow. 1976: Sold to R.B. Leasing Co, Edinburgh (Stirling Shipping Co Ltd). 1984: Sold to Royal Bank Leasing Co, Edinburgh (Stirling Shipping Co Ltd). 1986: Sold to British Linen Leasing Ltd, Edinburgh (Seadive Ltd, Jersey). Registered at Glasgow as SEA MUSSEL. Converted to a diving support vessel - 856gt, 257net. 1989: Glasgow registry closed. Registered at Hamilton, Bermuda. 05.1993: Sold to Putford Enterprises Ltd, Lowestoft. Hamilton registry closed. 06.05.1993: Registered at Lowestoft.
10.06.1993: Registered at Lowestoft as PUTFORD SEA MUSSEL. Converted to offshore platform standby safety vessel. 19.10.2002: Sold to Al Dowari General Trading Ajman, Kuwait. Lowestoft registry closed. Registered at Panama. 23.01.2003: Sold to Al-Mojil Group, Damman, Saudi Arabia. 01.2003: Lowestoft registry closed. Registered at Panama as AL MOJIL 41. 05.2003: Panama registry closed. Registered at Dammam. 07.2003: Remeasured to 825gt, 247net, 923dwt. 25.05.2011: Laid up.

STIRLING OAK	1554	699	51,01	W H Allen	Stirling Shipping Co Ltd,
361645 IMO 7342249	25.04.1974	299	11,61	2500bhp 2 x 8-cyl	Glasgow
TM supply ship	04.10.1974	923	4,65	12.5 knots	

Built by Cochrane & Sons Ltd, Selby (Drypool Group Ltd).
10.1974: Registered at Glasgow. 1976: Sold to R.B. Leasing Co, Edinburgh (Stirling Shipping Co Ltd). 1982: Sold to Seahorse Ltd, Cork, Co Cork. 1982: Sold to James Scott & Co (Cork) Ltd, Cork, Co Cork. Glasgow registry closed. Registered at Cork as SEAHORSE SUPPLIER. 21.09.2004: Sold to Raj Shipping Agencies Ltd, Mumbai, India. 09.2004: Registered at Mumbai as OFFSHORE SUPPLIER. 21.10.2014: Registry closed - "Broken up".

STIRLING EAGLE	1555	699	51,01	W H Allen	Stirling Shipping Co Ltd,
364965 IMO 7342251	03.09.1974	299	11,61	2500bhp 2 x 8-cyl	Glasgow
TM supply ship	22.01.1975	923	4,65	12.5 knots	

Built by Cochrane & Sons Ltd, Selby (Drypool Group Ltd).
12 passengers. 01.1975: Registered at Glasgow. 1976: Sold to R.B. Leasing Co, Edinburgh (Stirling Shipping Co Ltd). 1983: Sold to Royal Bank Leasing Co, Edinburgh (Stirling Shipping Co Ltd). 1987: Sold to Harrisons (Clyde) Ltd, Glasgow. Chartered to George Craig & Sons Ltd, Aberdeen. Glasgow registry closed. Registered at Aberdeen as GRAMPIAN EAGLE. 1990: Sold to George Craig & Sons Ltd, Aberdeen. Converted to offshore platform standby safety vessel. Remeasured to 805gt, 241net, 923dwt. 19.02.2003: Sold to Al-Mojil Group, Damman, Saudi Arabia. Aberdeen registry closed. 02.2003: Registered at Panama as AL MOJIL 44. With some modification operating as offshore supply vessel. 2003: Panama registry closed. Registered at Damman. 05.2004: Remeasured to 804gt, 241net, 923dwt. 29.09.2011: Laid up.

SEAFORTH WARRIOR	1556	802	51,77	British Polar	Seaforth Maritime Ltd,
359092 IMO 7342263	03.12.1974	309	11,80	6160bhp 2 x 16-cyl	Aberdeen
TM anchor handling	14.03.1975	1017	5,21	13.5 knots	
tug supply ship					

Built by Drypool Group Ltd, Selby; (Cochrane shipyard, Selby).
12 passengers. 03.1975: Registered at Aberdeen. 1976: Sold to R.B. Leasing Co Ltd, Edinburgh (Seaforth Maritime Ltd). 1977: Sold to "Brodospas" Poduzece za Spasavanje i Teglenje Brodova, Split, Yugoslavia. Aberdeen registry closed. Registered at Split as BRODOSPAS 21. 1991: Sold to "Brodospas" Poduzece za Offshore, Teglenje i Spasavanji, Split. Croatia. 1992: Registered at Split as BRODOSPAS WARRIOR. 05.1998: Registered at Split as BRODOSPAS JUGO. 2015: Believed operating as a private ship.

SEAFORTH VICTOR	1557	802	51,77	British Polar	Seaforth Maritime Ltd,
359093 IMO 7342677	30.08.1974	309	11,80	6160bhp 2 x 16-cyl	Aberdeen
TM anchor handling	27.03.1975		5,21	13.5 knots	
tug supply ship					

Sub contracted to and built by Charles Hill & Sons Ltd, Bristol (yard no.475).
12 passengers. 27.03.1975: Registered at Aberdeen. 1976: Sold to R.B. Leasing Co Ltd, Edinburgh. 1977: Sold to "Brodospas" Poduzece za Spasavanje i Teglenje Brodova, Split, Yugoslavia. Aberdeen registry closed. Registered at Split as BRODOSPAS 31. 1991: Sold to "Brodospas" Poduzece za Offshore, Teglenje i Spasavanji, Split. Croatia. 1992: Registered at Split as BRODOSPAS VICTOR. 05.1998: Registered at Split as BRODOSPAS BURA. 1999: Registered at Split as BRODOSPAS VICTOR. No trace.

SEAFORTH SAGA 359095 IMO 7340825 TM anchor handling tug supply ship	1558 24.02.1975 03.07.1975	802 309 1030	51,77 11,80 5,21	British Polar 6160bhp 2 x 16-cyl 13.5 knots	Seaforth Maritime Ltd, Aberdeen

Built by Drypool Group Ltd, Selby; (Cochrane shipyard, Selby).
12 passengers. 24.06.1975: Registered at Aberdeen. 1976: Sold to Lloyds & Scottish Development Ltd, Aberdeen. 03.1983: Sold to UK Ministry of Defence, Navy Department. Aberdeen registry closed. Converted to a patrol ship (2x40mm, AA weapons) by Commercial Drydocks Ltd, Cardiff. Commissioned as HMS **PROTECTOR** (P.No.P246). 27.10.1986: Returned from Falklands Islands. De-commissioned and placed on disposal list. 04.1987: Sold to Pounds Marine Shipping Ltd, Portsmouth, and offered for re-sale. 1990: Sold to Sociedad Colombiana de Servicios Portuarios S.A, Panama. Registered at San Lorenzo, Honduras, as **MOENA**. 20.07.1990: Sold to Serviport, Cartagena de Indias, Colombia. 01.09.1990: Registered at San Lorenzo as **SERVIPORT II**. Converted to FiFi tug supply ship. Remeasured to 776gt, 144net, 1040dwt. 09.1992: Registered at Cartagena de Indias, Colombia. 1997: Tonnage amended to 871gt. 2015: Still in service.

SEAFORTH CHAMPION 359090 IMO 7340837 TM anchor handling tug supply ship	1559 23.08.1974 23.01.1975	802 309 1029	51,77 11,80 5,21	British Polar 6160bhp 2 x 16-cyl 13.5 knots	Seaforth Maritime Ltd, Aberdeen

Built by Beverley S.B. & E. Co Ltd, Beverley.
12 passengers. 27.12.1974: Registered at Aberdeen. 1976: Sold to Lloyds & Scottish Development Ltd, Aberdeen. 03.1983: Sold to UK Ministry of Defence, Navy Department. Aberdeen registry closed. Converted to a patrol ship (2x40mm, AA weapons) by Commercial Drydocks Ltd, Cardiff. Commissioned as HMS **GUARDIAN** (P.No.P245). 1987: De-commissioned and placed on disposal list. 04.1987: Sold to Pounds Marine Shipping Ltd, Portsmouth, and offered for re-sale. 08.1987: Sold to associates of Mr V Cordeira, Lisbon. 03.1989: Vessel damaged by stranding. 1991: Sold to Dreyfus Naviera, La Guaira, Edo Vargas, Venezuela. Converted to FiFi tug supply ship. Registered at La Guaira as **GUARDIAN**. 2015: Still in service.

STELLAMAN 366045 IMO 7342275 Motor chemical tanker	1560 27.03.1975 02.1976	1513 835 2324	74,00 13,50 6,10	W H Allen 2600bhp 16-cyl 13.75 knots	Ingram Ocean Carriers Ltd, London

Built by Cochrane & Sons Ltd, Selby (Drypool Group Ltd).
02.1976: Registered at London. 1976: Sold to Transnational Insurance Ltd, London. 1979: Sold to C Rowbotham & Sons (Management) Ltd, London. 1981: Sold to Rowbotham Tankships Ltd, London. 1987: Sold to Lenitz Ltd, Hong Kong. London registry closed. Registered at Hong Kong as **CEDARWOOD**. 1989: Sold to Cedarwood Tankers Ltd, Hong Kong. 06.1989: Sold to Stolt Cedarwood Inc (Stolt-Nielsens Rederi A/S, Haugesund, Norway). Registered at Haugesund (NIS) as **STOLT CEDARWOOD**. 1990: Haugesund (NIS) registry closed. Registered at Monrovia, Liberia. 03.1995: Sold to Trans Ka Tanker Management Co Ltd, (Trans Ka Tanker Isletmeciligi Ticaret Ltd Sirketi), Istanbul, Turkey. Monrovia registry closed. Registered at Istanbul as **KEREM KA**. 07.2001: Sold to Granmar Denizcili Sanayive ve Ticaret A, Istanbul. Registered at Istanbul as **KARLITO**. 07.2006: Sold to Alriahe Shipping Co SA, Piraeus, Greece (part of Meander Mediterranean Shipping (Overseas) Co, SA, Piraeus). Istanbul registry closed. Registered at Moroni, Comoros, as **DIANA Z**. 04.2008: Sold to Crystallo NE, Atthens. 08.2008: Moroni registry closed. Registered at Piraeus as **AGIA ZONI I**. 2015: Still in service.

MARSMAN 366284 IMO 7392256 Motor chemical tanker	1561 22.08.1975 09.1976	1513 835 2324	74,00 13,50 6,10	W H Allen 2680bhp 16-cyl 13.75 knots	Ingram Ocean Carriers Ltd, London

Built by Cochrane & Sons Ltd, Selby (Drypool Group Ltd).
09.1977: Registered at London. 1977: Sold to Transnational Insurance Ltd, London. 1979: Sold to C Rowbotham & Sons (Management) Ltd, London. 1981: Sold to Rowbotham Tankships Ltd, London. 1987 Sold to Rimini Ltd, Hong Kong. London registry closed. Registered at Hong Kong as **BIRCHWOOD**. 1989: Sold to Cedarwood Tankers Ltd, Hong Kong. 06.1989: Sold to Stolt Birchwood Inc (Stolt-Nielsens Rederi A/S, Haugesund, Norway). Registered at Haugesund (NIS) as **STOLT BIRCHWOOD**. 06.1990: Haugesund (NIS) registry closed. Registered at Monrovia, Liberia. 05.1994: Sold to Karimata Investments Corporation, c/o PT Submare Laboro, Jakarta, Indonesia. Monrovia registry closed. Registered at Panama as **OCEEANA KAREEMATA**. 2015: Believed still in service.

TILSTONE MAID 338115 IMO 7392244 Motor cargo ship	1562 31.12.1974 25.03.1975	800 435 1184	57,61 10,20 4,58	W H Allen 1200bhp 8-cyl 12.0 knots	Tiling Construction Co Ltd, London

Built by Beverley S.B. & E. Co Ltd, Beverley.
Launched as **TILSTONE MAID**. Purchased on completion by Stag Line Ltd, North Shields. 03.1975: Start of 4-year bareboat charter to Silloth Shipping Ltd, Annan. Registered at North Shields as **SILLOTH STAG**. 14.12.1979: On loaded passage Inverkeithing towards Dover with stone, experienced engine problems off the Farne Islands. Tug connected and delivered Tyne Dock Engineering Ltd, South Shields, for repair. 24.11.1980: On loaded passage from Boston, Lincs, towards Belfast with feed wheat, sustained engine failure 34 miles off Pendeen Light, Cornwall; repaired by crew and proceeded on passage. 1983: Sold to Westfield Shipping Co Ltd, Barrow-in-Furness. 04.1985: Sold to J R Rix & Sons Ltd, Hull Shipping Co Ltd, Hull. 1985: Sold to Robrix Shipping Co Ltd, Hull. North Shields registry closed. Registered at Hull as **ROBRIX**. 03.12.1987: Entering Newlyn in heavy weather to load stone experienced engine failure and landed heavily on North Pier. Engine restarted and vessel berthed. 05.01.1990: On loaded passage from Dean Quarry towards London with roadstone experienced rudder problems. Company vessel JEMRIX (843grt/1965) connected and delivered to Falmouth. 1991: Remeasured to 798gt, 435net, 1184dwt. 1994: Hull registry closed. Registered at Nassau, Bahamas. 01.1996: Sold to Aned Maritime Ltd, c/o Seaflight Management Consultants Ltd, Dulwich, London. Nassau registry closed. Registered at Valletta as **SPRITE**. 16.09.1997: At Gibraltar with gearbox problems. 1997: Sold to Nortrans Shipping Group Inc, Belize. Valletta registry closed. Registered at Belize as **KONVIK**. 1998: Sold to Edarte Sh PK Import Export, Durres, Albania. Belize registry closed. 11.1998: Registered at Durres as **EDARTE**. 11.2002: Registered at Durres as **FROJDI I**. 16.03.2004: Sold to Alb Sea Transport Sh PK, Durres. 2015: Still in service.

The **Seaforth Saga** (1558)

The **Stellaman** (1560) was delivered to the Ingram Corporation.

ARMANA	1563	393	33,41	Mirrlees Blackstone	J Marr & Son Ltd,
365782 IMO 7366192	09.10.1975	156	8,50	1700bhp 16-cyl	Fleetwood
FD322 Motor stern trawler	06.05.1976		4,66	CPP 11.5 knots	

Built by Drypool Engineering and Dry Dock Co Ltd, Hull.
12.05.1976: Registered at Fleetwood (FD322).   07.1982: All eight of the company's remaining Fleetwood-based trawlers transferred to fish out of Hull.   1985: Under Sk Malcolm Trott, first white fish trawler to gross over £1,000,000 in a calendar year. Fishing home waters landed a total of 30,059 kits from 22 trips - 320 days at sea. Landing 9 trips at Bremerhaven & Cuxhaven, 6 at Aberdeen, 3 at Fleetwood, 2 at Birkenhead and 2 at Hull.   06.07.1987: Sold to Armana Ltd Hull.   01.12.1988: RSS No. A16738 allotted.   06.1994: Sold to Donegal Deep Sea Fishing Co Ltd, Killybegs, Co Donegal. Fleetwood registry closed. Registered at Dublin (D327).   09.04.1996: Sold to D Cook Ltd, New Holland, Lincolnshire, and broken up on the New Holland slipway. Dublin registry closed.

NAVENA	1564	393	33,41	Mirrlees Blackstone	J Marr & Son Ltd,
365784 IMO 7366207	05.12.1975	156	8,50	1700bhp 16-cyl	Fleetwood
FD323 Motor stern trawler	25.05.1976		4,66	CPP 11.5 knots	

Built by Beverley Shipbuilding & Engineering Co Ltd, Beverley.
02.07.1976: Registered at Fleetwood (FD323).   1977: Top Fleetwood trawler (Sk Victor Buschini), pairing with ARMANA (see yard no.1563), 19 trips grossed £470,284.   07.1982: All eight of the company's remaining Fleetwood-based trawlers transferred to fish out of Hull.   26.01.1984: In severe weather on passage Aberdeen to Hull (Sk Jeffrey Sumner); eleven crew. Off Flamborough Head started to take in water forward and developed a starboard list. Fearing that the trawler would capsize a distress call was sent and crew look to liferafts but skipper remained onboard to direct the rescue. Liferaft capsized and all crew picked up from water by RAF helicopter. Sk Sumner later picked up by helicopter. Hull tug YORKSHIREMAN (see yard no.104) engaged to take trawler in tow but in the meantime three Scarborough fishing vessels connected and CASSAMANDA (SH128) (110grt/1969) commenced tow to Scarborough.   27.01.1984: Having been refused entry to harbour and directed to beach the vessel, put ashore south of the harbour and later capsized.   05.1984: Attempts to refloat failed and Scarborough Council took possession of the wreck.   23.05.1984: Fleetwood registry closed - "Total Loss".   Subsequently salved. Sold to Marine Partners Ltd, Middlesbrough.   03.1985: Sold to S & S Sunderland Marine Ltd, Sunderland, in damaged condition for repair. 1991: Repair considered to be uneconomical.   Broken up at Sunderland.

BISHOP BURTON	1565	165	22,99	Caterpillar	Mike Burton Ltd,
365577 IMO 7400390	25.04.1975	60	7,29	850bhp 12-cyl	Hull
H293 Motor stern trawler	31.07.1975		3,59	9.0 knots	

Built by Beverley Shipbuilding & Engineering Co Ltd, Beverley.
1975: Registered at Hull (H293).   1980: Transferred to fish out of North Shields.   29.01.1985: Foundered some 26 miles off Aberdeen in approximate position 56.51N 01.27W. All crew taken off by helicopter.   Hull registry closed.

STIRLING ASH	1566	699	51,01	W H Allen	Stirling Shipping Co Ltd,
365001 IMO 7400405	26.08.1975	299	11,61	2400bhp 2 x 8-cyl	Glasgow
TM supply ship	16.03.1976	913	4,65	12.5 knots	

Built by Beverley S.B. & E Co Ltd, Beverley.
12 passengers.   03.1976: Registered at Glasgow.   1986: Sold to Seadive Ltd, Jersey, Channel Islands. Registered at Glasgow as SEA OYSTER. Converted to support survey & diving operations.   11.1991: Sold to George Craig & Sons Ltd, Aberdeen. Converted to offshore platform stand-by safety vessel. Glasgow registry closed. Registered at Aberdeen.   08.1992: Registered at Aberdeen as GRAMPIAN SABRE.   19.02.2003: Sold to Mohammad Al Mojil Group, Dammam, Saudi Arabia. Aberdeen registry closed. Registered at Panama as AL MOJIL 42.   10.2003: Panama registry closed.   10.2003: Registered at Dammam.   04.2004: Remeasured to 819gt, 913dwt. 25.05.2011: Laid up at Dammam.

STIRLING SWORD	1567	699	51,01	W H Allen	Stirling Shipping Co Ltd,
376961 IMO 7400417	26.08.1975	299	11,61	2400bhp 2 x 8-cyl	Glasgow
TM supply ship	16.03.1976	914	4,68	12.5 knots	

Built by Drypool Eng & Dry Dock Co Ltd, Hull.
010.1976: Registered at Glasgow.   1984: Sold to George Craig & Sons Ltd, Aberdeen. Converted to offshore platform standby safety vessel. Glasgow registry closed. Registered at Aberdeen as GRAMPIAN SWORD.   02.2003: Sold to Mohammad Al Mojil Group, Dammam, Saudi Arabia. Aberdeen registry closed. Registered at Panama as AL MOJIL 43.   10.2003: Panama registry closed. 10.2003: Registered at Dammam.   04.2004: Remeasured to 817gt, 914dwt.   29.09.2011: Laid up at Dammam.

EXEGARTH	1568	388	34,02	Ruston Paxman	Rea Towing Co Ltd,
376349 IMO 7400950	17.05.1976		10,32	3520bhp 16-cyl	Liverpool
Motor tug FiFi	20.10.1976		4,91	13.0 knots	

Built by Beverley Shipbuilding & Engineering Co Ltd, Beverley.
10.1976: Registered at Milford Haven.   1988: Sold to Irish Tugs Co Ltd, Westport, Co Mayo, Irish Republic.   1991: Milford Haven registry closed. Registered in Westport.   08.1995: Sold to Sysgold Business Inc, Panama. Westport registry closed. Registered at Panama as PAGBILAO I. Remeasured to 367gt, 110net.   05.1998: Sold to I.D.H.I. Ports & Shipping Corporation, Cagayan de Oro, Philippines. Panama registry closed. Registered at Cagayan de Oro.   08.1998: Sold to Pagbilao Shipping Corporation, Cebu City, Cebu, Philippines. 21.06.2012: Laid up.

Not built	1569 Order cancelled.

Armana (1563).

(Cochrane archive)

Bishop Burton (1565)

(Cochrane archive)

Exegarth (1568)

(Cochrane archive)

BOSTON SEA KING	1570	171	22,99	Mirrlees Blackstone	Boston Deep Sea
362290　IMO 7409308	29.07.1975	59	7,31	700bhp　8-cyl	Fisheries Ltd,
LT265　Motor stern trawler	09.01.1976		3,51	CPP　10.0 knots	Hull

Hull built by J R Hepworth & Co (Hull) Ltd, Hull (yard no.119), completed at Selby.
12.01.1976: Registered at Lowestoft (LT256).　1983: Sold to Natal Inshore Trawlers (Pty), Durban, South Africa.　20.06.1983: Lowestoft registry closed.　Registered at Durban as **SEA KING** (DNA52).　1988: Sold to Natal Ocean Trawling (Pty), Durban.　Registered at Durban as **OCEAN KING** (DNA52).　1995: Sold to Viking Fishing Co (Pty) Ltd, Durban.　08.1997: In Durban in poor condition.　26.03.2003: In Cape Town.　No further trace.

BOSTON SEA KNIGHT	1571	171	22,99	Mirrlees Blackstone	Boston Deep Sea
376314　IMO 7409310	03.02.1976	59	7,31	700bhp　8-cyl	Fisheries Ltd,
LT319　Motor stern trawler	28.05.1976		3,51	CPP　10.0 knots	Hull

Built by J R Hepworth & Co (Hull) Ltd, Hull (yard no.120).
02.06.1976: Registered at Lowestoft (LT319).　1984: Employed as offshore standby safety vessel.　01.03.1984: Sold to W H Kerr (Ship Chandlers) Ltd, Milford Haven.　21.12.1987: Sold to Britannia Marine Ltd, Lowestoft. Converted to an offshore platform standby safety vessel.　Registered at Lowestoft as **BRITANNIA KNIGHT**.　01.1995: Sold to Cove Beech Ltd, Lowestoft.　Registered at Lowestoft as **COVEX BRILLIANT**.　01.2003: Sold to Malta.　Converted for tuna fishing/farming.　Remeasured to 146gt, 91net.　Lowestoft registry closed. Registered at Valletta as **ROSARIA TUNA** (MFA511).　2015: Still in service.

BOSTON SEA RANGER	1572	171	23,14	Mirrlees Blackstone	Boston Deep Sea
376317　IMO 7409322	26.08.1976	59	7,31	700bhp　8-cyl	Fisheries Ltd,
LT328　Motor stern trawler	01.12.1976		3,61	CPP　10.0 knots	Hull

Hull built by J R Hepworth & Co (Hull) Ltd, Hull (yard no.121), completed at Selby.
01.12.1976: Registered at Lowestoft (LT328).　05.12.1977: Fishing mackerel off the Cornish coast (Sk Ian Lace); eight crew. Two miles off Gwennap Head, overwhelmed by a following sea and with fishroom hatches open, flooded fishroom. Crew abandoned and shortly after capsized and foundered in approximate position 49.59N 05.39W. Five of her eight crew were lost; survivors including skipper picked up by Sennon Cove lifeboat.　08.12.1978: Lowestoft registry closed.　07.1979: At the formal investigation held at Gt Yarmouth (S501), the Court found that the loss of the BOSTON SEA RANGER was partly caused or contributed to by the wrongful act or default of her skipper, and partly caused or contributed to by the wrongful act or default of her owners.

SEAFORTH CONQUEROR	1573	1432	58,50	Mirrlees Blackstone	Seaforth Maritime
359113　IMO 7406019	15.06.1976	609	13,71	7320bhp　4 x 12-cyl	(Conqueror) Ltd,
TM tug supply ship	30.12.1976	1236	6,76	CPP　14.3 knots	Aberdeen

Built by Cochrane Shipbuilders Ltd, Selby.
30.12.1976: Registered at Aberdeen.　1988: Sold to Boa Ltd c/o Taubatkompaniet A/B, Trondheim, Norway. Aberdeen registry closed. Registered at Georgetown, Cayman Islands, as **BOA CONQUEROR**. Operating as an anchor handler/supply ship.　02.1995: Sold to Toisa Ltd c/o Sealion Shipping Ltd, Farnham, Surrey.　Registered at Georgetown as **TOISA CONQUEROR**.　Fitted out FiFi.　1999: Sold to Demeresa-Desguaces Y Relaminables, Mexico.　26.10.1999: Arrived Tuxpan for breaking up.　Registry closed.

SEAFORTH CLANSMAN	1574	1977	68,71	Mirrlees Blackstone	Seaforth Maritime Ltd,
377895　IMO 7406021	08.03.1977	577	13,71	7320bhp　4 x 12-cyl	Aberdeen
TM diving support ship	30.07.1977	1180	6,76	CPP　13.0 knots	

Built by Cochrane Shipbuilders Ltd, Selby.　Also equipped for fire fighting and pollution control.
08.1977: Registered at Aberdeen.　1979: Sold to Seaforth Maritime (Clansman) Ltd, Log Ships Ltd, and Prentec Ltd, Aberdeen.
1981: Chartered by Ministry of Defence (Navy) London (civilian crew) to accommodate Naval Party 1007 as a deep sea saturation diving platform following decommissioning of HMS RECLAIM (P.No.A23)) and to await commissioning of HMS CHALLENGER (P.No.K07).
19.07.1983: Recovered fuselage of Sikorsky S.61N helicopter which had crashed in the sea off St Mary's Airport, Isles of Scilly, when on a flight from Penzance to Scilly. Twenty passengers died and six survivors were picked up by St Mary's lifeboat ROBERT EDGAR.
1987: Charter completed.　1988: Sold to Commissioners of Irish Lights, Dublin. Aberdeen registry closed. Registered at Dublin as **GRAY SEAL**. Converted to buoy and lighthouse tender.　03.1994: Sold to James Fisher & Sons, (Liverpool) Ltd, Liverpool. Converted to a survey vessel.　01.1995: Transferred to Coe Metcalf Shipping Ltd, Liverpool.　Registered at Dublin as **OSV ZEALOUS**.　06.1995: Transferred to James Fisher & Sons PLC, Barrow-in-Furness.　Dublin registry closed. Registered at Barrow.　08.1995: Sold to De Beers Marine (Pty) Ltd, Cape Town, South Africa.　1996: Barrow registry closed. Registered at Cape Town as **ZEALOUS**.　Remeasured to 2285gt, 1090net, 1206dwt.　2001: Sold to Miliana Shipping Co Ltd, c/o Miliana Shipmanagement Ltd, Nicosia, Cyprus.　Cape Town registry closed. Registered at Kingstown, St. Vincent & the Grenadines as **ST. BARBARA**.　07.2001: Converted to geophysical research ship. 06.2013: Following update remeasured to 2324gt, 1090net, 1180dwt.　2015: Still in service.

All ships now built by Cochrane Shipbuilders Ltd, Selby

LADY MOIRA	101	348	31,02	Ruston Diesels	United Towing (Yeoman) Ltd,
376885　IMO 7610012	07.04.1977		10,01	2920bhp　2 x 8-cyl	Hull
TM tug	01.11.1977		5,11	11.5 knots	

09.11.1977: Registered at Hull.　10.04.1978: Sold to United Towing (Nelson) Ltd Hull.　15.02.1979: In storm force winds, snow and heavy seas went to assistance of Romanian motor vessel SAVINESTI (2473grt/1978) in danger of being pushed on the banks off the Norfolk coast. Weather conditions too severe to allow men on deck. Stood by.　16.02.1979: As weather improved escorted SAVINESTI to River Humber.
14.04.1983: Sold to Humber Tugs (Nelson) Ltd, Grimsby.　24.03.1988: Sold to Humber Tugs Ltd, Grimsby.　06.06.1992: With LADY ANYA (364grt/1990), attending a tanker at Tetney mono buoy, gog rope parted and pulled over broadside. Quick release operated but water entered engine room via starboard funnel and one man in water. Man picked up by Humber lifeboat KENNETH THELWELL plus one other crewman transferred. LADY ANYA connected and delivered Immingham.　31.01.1994: Remeasured to 369gt, 110net.　1996: Sold to Howard Smith (Humber) Ltd, Hull.　05.1997: Sold to Multratug BV, Terneuzen, Netherlands, c/o Handel-en Scheepvaartmaatschappij Multraship BV, Terneuzen.　Hull registry closed. Registered at Terneuzen as **MULTRATUG 7**.　02.2011: Chartered by MTS Group Ltd, Falmouth. 02.2011: Registered at Terneuzen as **MTS VISCOUNT**.　2015: Still in service.

LADY MOIRA - FROM START TO SERVICE

Immediately after launch.

(Cochrane archive)

Fitting out.

(Bernard McCall)

In service with Humber Tugs and about to enter the lock at Immingham.

(Bernard McCall)

LADY DEBBIE 376899 IMO 7610024 TM tug	102 18.08.1977 24.07.1978	348	31,02 10,01 5,11	Ruston Diesels 2920bhp 2 x 8-cyl 11.5 knots	United Towing (Exmouth) Ltd, Hull	

04.08.1978: Registered at Hull.   14.04.1983: Sold to Humber Tugs (Exmouth) Ltd, Grimsby.   24.03.1988: Sold to Humber Tugs Ltd, Grimsby.   31.01.1994: Remeasured to 369 gt, 110 net.   1995: Sold to Howard Smith (Humber) Ltd Hull.   05.01.1998: Went to the assistance of the disabled Liberian chemical tanker MULTITANK BAHIA (3726gt/1996) drifting in the North Sea with engine failure. In approx position 28.07N 00.53E connected and delivered Immingham the same day.   07.2003: Sold to Adsteam Humber Ltd, Hull (part of Adsteam Marine Group, Sydney).   10.2007: Sold to Svitzer Humber Ltd, Immingham, (part of Svitzer AS, Copenhagen).   21.10.2009: Sold to Somara, Fort de France, Martinique.   10.2013: Hull registry closed.   11.2013: Registered at Kingstown, St Vincent and the Grenadines. 2015: Still in service.

IRISHMAN 376889 IMO 7621499 TM tug	103 11.01.1978 14.03.1978	641 106	36,50 11,00 5,74	Ruston Diesels 4380bhp 2 x 12-cyl CPP 13.0 knots	United Towing (Drake) Ltd, Hull	

30.03.1978: Registered at Hull.   19.06.1980: Remeasured to 685grt, 72net.   17.04.1982: Requisitioned under Ships Taken Up From Trade (STUFT) system by Department of Trade, Shipping Policy Division, for service in the South Atlantic during the Falklands campaign (Capt Tony Allen, master).   25.05.1982: Off the Falkland Islands the STUFT RoRo/container vessel ATLANTIC CONVEYOR (14,950gt /1970) fitted out as a supply ship/aircraft ferry was hit by two Argentine air-launched AM39 exocet missiles. Went to the assistance of the burning vessel, connected and commenced tow but burned out hulk started to settle.   28.05.1982: Tow released and foundered. 1984-1986: Chartered by Ministry of Defence (Navy) for service in the South Atlantic.   08.09.1988: Sold to Sembawang Salvage (IV) Pte Ltd, Singapore. Registered at Hull as **SALVISION**.   16.01.1989: Hull registry closed. Registered at Singapore.   03.1993: Sold to Pacnav S.A, Panama. Singapore registry closed. Registered at Panama as **PACNAV ACE**.   13.08.1993: Sold to Armamex Naviera S.A de C.V, Mexico City, Mexico.   12.1993: Panama registry closed. Registered at Ciudad del Carmen as **EL HUASTECO**.   2015: Still in service.

YORKSHIREMAN 376895 IMO 7621504 TM tug	104 28.03.1978 24.06.1978	686	36,50 11,00 5,74	Ruston Diesels 4380bhp 2 x 12-cyl CPP 13.0 knots	United Towing (Tradesman) Ltd, Hull	

Completed for United Towing (Hood) Ltd, Hull.   03.06.1978: Registered at Hull.   17.04.1982: Requisitioned under Ships Taken Up From Trade (STUFT) system by Department of Trade, Shipping Policy Division, for service in the South Atlantic during the Falklands campagn (Capt Peter Rimmer, master).   26.01.1984: Stern trawler NAVENA (see yard no.1564), when off Flamborough Head on passage Aberdeen towards Hull, started to take in water and developed a startboard list. Engaged to tow but stood down when local fishing boats connected and commenced tow to Scarborough.   1984-1986: Chartered by Ministry of Defence (Navy) for service in the South Atlantic. 09.12.1986: Hull registry provisionally closed. Registered at Nassau, Bahamas.   03.06.1987: Hull registry finally closed on discharge of mortgage.   1988: Sold to Sembawang Salvage (IV) Pte Ltd, Singapore. Bahamas registry closed. Registered at Singapore as **SALVIGOUR**.   1989: Sold to Slibail Portuguesa de Locação Financeira S, Lisbon, Portugal. Singapore registry closed. Registered at Lisbon as **COMENDA**.   01.1995: Sold to REBOSADO – Rebocadores Fluviais do Sado Ltda, Setubal.   14.10.2005: Sold to Jens Alfastsen Rederiet Stenderup, Denmark. Lisbon registry closed. Registered at Horsens as **MIRA A**.   23.02.2007: Sold to Sanpaola Leasint S p a-Societa di Leasing Internazionale, Milan, and leased to Rimorchiatori Meridionali, Srl, Naples, Italy. Horsens registry closed. 02.2007: Registered at Naples as **MARECHIARO**.   2009: Sold to Rimorchiatori Meridionali, Srl, Naples.   22.07.2014: Sold to International Ships Trading, Panama City. Naples registry closed.   07.2014: Registered at Panama.   2015: Still in service.

BROMLEY 379715 IMO 7711103 Motor oil tanker	105 23.06.1978 20.09.1978	640 343 844	53,52 10,61 3,61	Alpha-Diesel 1240bhp 8-cyl CPP 10.5 knots	Bowker & King Ltd, London	

09.1978: Registered at London.   1979: Sold to Nile S.S Co Ltd, London. Managed by Bowker & King.   1984: Sold to Investors in Industry P.L.C, London. Managed by Bowker & King.   1989: Company restyled 3i's plc, London. Managed by Bowker & King.   10.1993: Sold to John H Whitaker (Tankers) Ltd, Hull. Registered at London as **WHITASK**.   06.2009: Sold to Nature Group PLC, St Helier, Jersey. 06.2009: London registry closed. Registered at Gibraltar as **HUMBER MIST**.   11.06.2012: Laid up Gibraltar.

SELBYDYKE 386446 IMO 7711103 TM diving support ship	106 16.11.1978 08.02.1979	1598 1066	74,02 13,20 6,00	Mirrlees Blackstone 3000bhp 6-cyl 13.0 knots	Klondyke Shipping Co Ltd, Hull	

09.02.1979: Registered at Hull.   08.07.1982: Company restyled North British Shipping Co Ltd, Hull.   02.03.1985: Sold to Primrose Fishing Co Ltd Hull.   11.1985: Sold to Norbrit Waal B.V, Rotterdam, Holland.   08.11.1985: Hull registry closed. Registered at Rotterdam as **NORBRIT WAAL**.   1987: Sold to Santa Monica Ltd, Nassau, Bahamas. Rotterdam registry closed. Registered at Nassau.   05.1988: Sold to Bjorgun H/F, Reykjavik, Iceland. Nassau registry closed. Converted to a suction hopper dredger. 1705gt, 512net. Registered at Reykjavik as **SOLEY**.   2015: Still in service.

STAR PERSEUS 377907 IMO 7803322 TM diving support ship	107 14.03.1979 16.05.1979	492 156 396	38,26 9,52 4,27	Caterpillar 1330bhp 2 x 8-cyl CPP 12.0 knots	Star Offshore Services Marine Ltd, Aberdeen	

05.1979: Registered at Aberdeen.   1980: Sold to Star Offshore Services Marine Ltd, Aberdeen.   1988: Sold to the Government of New Zealand (Ministry of Defence), Wellington, for the sum of NZ$1.5 million. Aberdeen registry closed.   05.04.1988: Commissioned as HMNZS **MANAWANUI** and sailed for New Zealand still in Star Offshore colours. On arrival converted for diving and mine counter measures support (P.No.A09).   Later further converted as a base ship for the Littoral Warfare Support Force.   2015: Still in service.

Fitting out of **Lady Debbie** (102) appears to be almost complete and delivery is imminent.

(Bernard McCall)

Irishman (103)

(Cochrane archive)

Star Perseus (107)

(Cochrane archive)

BOSTON SEA LANCE 386483 IMO 7813925 Motor refrigerated cargo ship	108 12.07.1979 25.09.1979	696/1463 371/812 925/1815	68,03 12,01 7,12	Mirrlees Blackstone 3000bhp 6-cyl 13.5 knots	Boston Deep Sea Fisheries Ltd, Hull

02.10.1979: Registered at Hull. 08.07.1982: Company restyled North British Shipping Co Ltd, Hull. 02.03.1985: Sold to Primrose Fishing Co Ltd, Hull. 11.1983: Sold to Norbrit Vries B.V., Rotterdam, Netherlands. 11.1983: Hull registry closed. 08.11.1983: Registered at Rotterdam as **NORBRIT VRIES**. 22.12.1983: Hull registry closed 1985: Lengthened; remeasured to 2004gt, 1815dwt. 1986: Sold to Santa Monica Ltd, Nassau, Bahamas. Rotterdam registry closed. 12.1986: Registered at Nassau. 1988: Sold to Sunset Shipping Co, Nassau. Registered at Nassau as **FENLAND**. 11.2005: Sold to JSC Arcticservice, Murmansk, Russia. 11.06.2007: Sold to Saly Shipping Co SA, Panama. Nassau registry closed. Registered at Panama as **SALY REEFER**. 10.2008: Sold to Fishing & Cargo Services SA, Panama. Panama registry closed. 03.11.2008: Registered at Moroni, Comoros. 2015: Still in service.

LIZZONIA 364574 IMO 7827327 Motor cargo ship	109 05.12.1979 04.02.1980	798 554 1315	56,01 11,21 4,63	Mirrlees Blackstone 999bhp 8-cyl 12.0 knots	J Wharton (Shipping) Ltd, Gunness

05.02.1980: Registered at Goole. 26.06.1986: Acquired by F T Everard Shipping Ltd, Greenhithe, following a de-merger deal. 13.02.1989: Registered at Goole as **CAPACITY**. 06.10.1993: Sold to Onesimus Dorey (Shipowners) Ltd. St.Peter Port, Guernsey, and demise chartered to Torbulk Ltd, Grimsby. 17.01.1994: Goole registry closed. 18.01.1994: Registered at Grimsby. 20.01.1994: Registered at Grimsby as **PENTLAND**. 1998: Sold to Pentland Marine Ltd, Grimsby. 03. 2000: Sold to Merlin Marine Ltd, Grimsby. 03.2000: Grimsby registry closed. Registered Bridgetown, Barbados, as **SEA KESTREL**. 03.11.2003: Sold to Faversham Ships Ltd, Faversham. 04.11.2003: Registered at Bridgetown as **ISLAY TRADER**. 05. 2005: Sold to Riga Navigation Co Ltd, St Kitts & Nevis. Registered at Bridgetown as **SEA STAR**. 06. 2006: Sold to Auriga Shipping Co Ltd, Monte Carlo, Monaco. 01.2007: Bridgetown registry closed. 02. 2007: Registered at Valletta, Malta, as **SAMOS ISLAND**. 02. 2007: Sold to Isola Services S.A, Piræus, Greece. 12.2010: Registered at Valletta as **SON I**. 19.01.2011: Sold to Son Shipping & Trading Ltd c/o Derin Shipping Agency & Shipping Management Trade Co Ltd, Istanbul. 2015: Still in service.

ANGELONIA 364575 IMO 7827249 Motor cargo ship	110 19.02.1980 01.04.1980	798 554 1315	56,01 11,21 4,63	Mirrlees Blackstone 990bhp 8-cyl 11.0 knots	J Wharton (Shipping) Ltd, Gunness

02.04.1980 Registered at Goole. 26.06.1986: Acquired by F T Everard Shipping Ltd, Greenhithe, following a de-merger deal. 07.06.1988: Registered at Goole as **COMITY**. 06.10.1993: Sold to Onesimus Dorey (Shipowners) Ltd, St Peter Port, Guernsey, and demise chartered to Torbulk Ltd, Grimsby. 19.10.1993: Goole registry closed. 20.10.1993: Registered at Grimsby. 29.10.1993: Registered at Grimsby as **PORTLAND**. 01.1997: Sold to Torbulk Ltd, Grimsby. 03.1997: Sold to John Fleming & Co (Holdings) Ltd, Aberdeen. Grimsby registry closed. 06.1997: Registered at Lerwick as **SHETLAND TRADER**. 22.04.2002: Sold to Faversham Ships Ltd, Faversham. Lerwick registry closed. Registered at Bridgetown, Barbados. 08.2007: Sold to Frakt & Sandsølan SP/F, Runavik, Faroe Islands. Bridgetown registry closed. Registered at Runavik as **HAV TIND**. Komatsu PC290LC-6 deck excavator fitted. 2015: Still in service.

ESSO PLYMOUTH 387642 IMO 7902300 Motor tanker	111 14.06.1980 11.09.1980	1421 821 2162	66,50 12,50 5,77	A P E Allen 2243bhp 6-cyl CPP 11.75 knots	Esso Petroleum Co Ltd, London

09.1980: Registered at Southampton. 01.1989: Sold to Rowbotham Tankships Ltd, London. Southampton registry closed. Registered at Douglas, Isle of Man, as **GUIDESMAN**. 1992: Sold to Rowbotham Tankships (Gibraltar) Ltd, Gibraltar. Douglas registry closed. Registered at Gibraltar. 01.02.1993: Sold to P&O Tankships (Gibraltar) Ltd, Gibraltar. 12.1994: Sold to Hui Quan Pte Ltd, Singapore. Gibraltar registry closed. Registered at Singapore as **BERJAYA DUA**. Remeasured to 1405gt, 739net, 2162dwt. 1995: Lengthened by Pan United Shipping Pte Ltd, Singapore. Remeasured to 75,50m, depth 5,75m. 1593gt, 938net, 2500dwt. 10.1999: Sold to Prestige Marine Services Pte Ltd, Singapore. 07.2004: Sold to Lisa Ltd, Wonsan, North Korea. Singapore registry closed. Registered at Wonsan as **AJMAN SUPPLIER**. 01.2005: Registered at Wonsan as **AJMAN GLORY**. 07.2006: Sold to Shipping World LLC, Dubai, United Arab Emirates. Wonsan registry closed. 07.2006: Registered at Moroni, Gran Comoros, as **BERJAYA DUA**. 21.06.2010: Registry closed - "Broken up".

URGENCE 389611 IMO 7928029 Motor cargo ship	112 11.11.1980 17.02.1981	699 435 1842	81,01 11,41 5,41	B & W Alpha 1240bhp 8-cyl 11.0 knots	Babyssa Ltd, Rochester

02.1981: Registered at Rochester. 1984: Sold to London & Rochester Trading Co Ltd, (Crescent Shipping), Rochester. Kent. 1987: Sold to Crescent Shipping Ltd, Rochester. 11.1992: Rochester registry closed. Registered at Nassau, Bahamas. 07.1994: Engine re-rated to 1013 bhp, 10 knots; remeasured to 1425gt, 497net. 11.1997: Sold to Crescent Navigation Ltd, Fareham. 2000: Sold to Crescent Tankships Ltd, Southampton. 2001: Laid up at Otterham Quay. 01.2002: Sold en bloc to clients of Arpa Shipping BV, Roosendaal, with STRIDENCE (see yard no.123); TURBULENCE (see yard no.124) and VIBRENCE (see yard no.113). 01.2002: Sold to Helen Ltd, c/o SMS Ship Management Support BV, Klundert, Netherlands. Registered at Nassau as **HELEN**. 2014: Sold to unidentified owners. Nassau registry closed. Registered at Aviatu, Cook Islands, as **URGENCE**. 2015: Still in service.

VIBRENCE 389626 IMO 7928031 Motor cargo ship	113 09.02.1981 29.04.1981	699 435 1842	81,01 11,41 5,41	B & W Alpha 1240bhp 8-cyl 10.5 knots	Babyssa Ltd, Rochester

04.1981: Registered at Rochester. 1982: Sold to London & Rochester Trading Co Ltd, (Crescent Shipping), Rochester. 1987: Sold to Crescent Shipping Ltd, Rochester. 11.1992: Rochester registry closed. Registered at Nassau, Bahamas. 07.1994: Engine re-rated to 1013 bhp, 10 knots; remeasured to 1425gt, 497net. 11.1997: Sold to Crescent Navigation Ltd, Fareham. 2000: Sold to Crescent Tankships Ltd, Southampton. 2001: Laid up at Otterham Quay. 01.2002: Sold en bloc to clients of Arpa Shipping BV, Roosendaal, with STRIDENCE (see yard no.123); TURBULENCE (see yard no.124) and URGENCE (see yard no.112). 01.2002: Sold to m.v. Laura Ltd c/o Arpa Shipping BV, Rotterdam. Registered at Nassau as **LAURA**. 2006: Sold to Laura Ltd c/o SMS Ship Management Support BV, Klundert, Netherlands. Registered at Nassau as **LAURA C**. 10.2008: Sold to Traverse Shipping Co Ltd c/o Arpa Shipping BV, Rotterdam. Registered at Nassau as **TRAVERSE**. 28.08.2011: Sold to Lumiere S Shipping Ltd c/o Dabmar Ship Management NV, Curaçao. 08.2011: Nassau registry closed. Registered at Panama as **SENA T**. 30.12.2013: Broken up. Registry closed.

General arrangement drawing
of **Boston Sea Lance** (108).

(Cochrane archive)

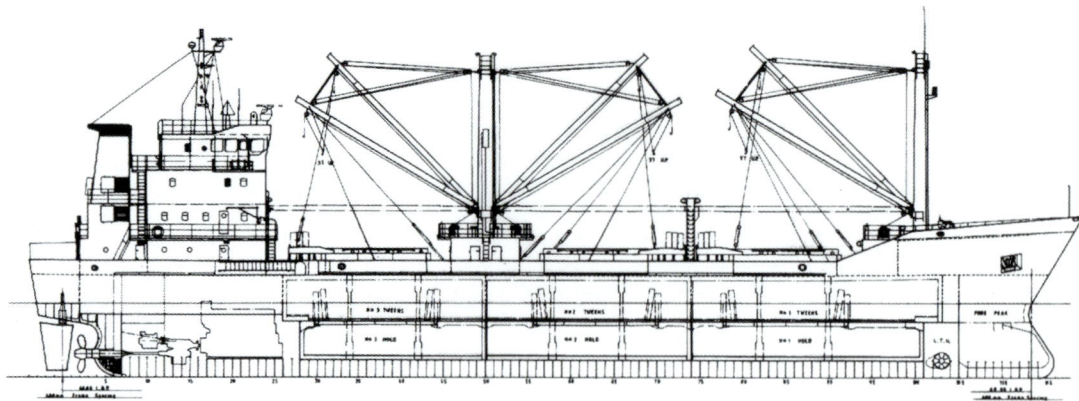

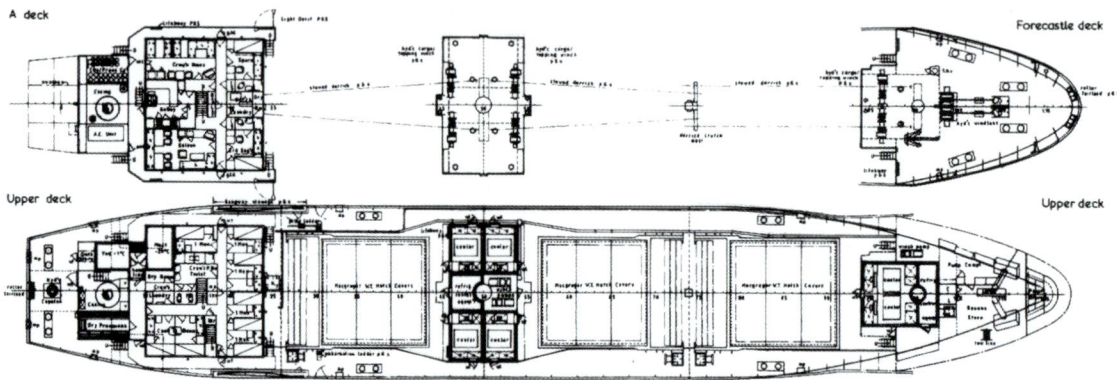

BOSTON SEA LANCE

GENERAL PARTICULARS

LENGTH OVERALL		73·976 m
LENGTH B.P.		68·000 m
BREADTH MLD		12·000 m
DEPTH MLD	Lower Deck	3·900 m
DEPTH MLD	Upper Deck	7·100 m
DRAFT Extreme	S.L.W.L.	5·214 m
GROSS TONNAGE	C.S.D.	1469·23
GROSS TONNAGE	O.S.D.	895·51
NETT TONNAGE	C.S.D.	817·81
NETT TONNAGE	O.S.D.	374·45
DEADWEIGHT		1815 Tonnes
CLASSIFICATION	LLOYDS ⊕ 100A1 ⊕ LMC	
	UMS ⊕ RMC ICE CLASS 3	
MAIN ENGINE	MIRRLEES BLACKSTONE KMR MAJOR	
POWER	3000 BHP at 500 R.P.M.	
TOTAL OIL FUEL		460·4 Tonnes
TOTAL BALLAST WATER		320·5 Tonnes
TOTAL FRESH WATER		54·0 Tonnes

SCALE 1:100

Boston Sea Lance during fitting out.

(Bernard McCall)

130

The launch of **Lady Elizabeth** *(114).*

(Frank Flintoft, Roy Cressey collection)

Lady Elizabeth *eases a BP tanker on to a jetty at Immingham in July 1981.*

(Bernard McCall)

LADY ELIZABETH 389062 IMO 8003644 Motor tractor tug	114 22.04.1981 02.07.1981	268	28,23 9,21 3,81	Ruston Diesels 2640bhp 2 x 6-cyl 2 x VSP 12.5 knots	Humber Tugs Ltd, Hull

12.1980: Keel laid.   08.07.1981: Registered at Hull.   01.08.1985: Transferred to operate out of Grimsby.   31.01.1994: Remeasured to 285gt, 85 net.   1995: Sold to Howard Smith (Humber) Ltd, Hull.   2003: Company restyled Adsteam Humber Ltd, Hull.   12.2005: Registered at Hull as **ADSTEAM ELIZABETH**.   08.2006: Sold to Humber Tugs Ltd, Hull.   Registered at Hull as **HT SABRE**.   10.2007: Sold to Svitzer Humber Ltd, Immingham.   10.2008: Registered at Hull as **SVITZER HUMBER**.   05.05.2009: Sold to T P Towage Co Ltd, Gibraltar. Registered at Hull as **ROOKE**.   2015: Still in service at Gibraltar.

LADY CONSTANCE 389072 IMO 8102141 Motor tractor tug	115 12.01.1982 08.03.1982	268	28,30 9,20 3,81	Ruston Diesels 2640bhp 2 x 6-cyl 2 x VSP 12.5 knots	Humber Tugs Ltd, Hull

12.03.1982: Registered at Hull.   01.08.1985: Transferred to operate out of Grimsby.   31.01.1994: Remeasured to 285 gross, 85 net. 1995: Sold to Howard Smith (Humber) Ltd, Grimsby.   2003: Company restyled Adsteam Humber Ltd, Hull.   10.2007: Sold to Svitzer Humber Ltd, Immingham.   07.2007: Registered at Hull as **SVITZER CONSTANCE**.   08.2014: Sold to Fowey Harbour Commissioners. Renamed **CANNIS**.   2015: Still in service.

LANMAR CREST 389190 IMO 8104254 TM tug/mooring tender; FiFi	116 06.07.1981 19.08.1981	133	22,81 6,61 3,43	Caterpillar 1450bhp 2 x 12-cyl 10.0 knots	Land and Marine Engineering Ltd, Bromborough

08.1981: Registered at Liverpool.   1993: Sold to Costain Oil, Gas & Process Ltd, Wythenshaw, Greater Manchester.   2006: Sold to Forcados Lineboats Ltd for use at Forcados Oil Terminal, Niger Delta, Nigeria.   2015: Still in service.

LANMAR REACH 389193 IMO 8104266 TM tug/mooring tender; FiFi	117 20.08.1981 29.09.1981	133	22,81 6,61 3,43	Caterpillar 1450bhp 2 x 12-cyl 10.0 knots	Land and Marine Engineering Ltd, Bromborough

10.1981: Registered at Liverpool.   1993: Sold to Costain Oil, Gas & Process Ltd, Wythenshaw, Greater Manchester.   2006: Sold to Forcados Lineboats Ltd for use at Forcados Oil Terminal, Niger Delta, Nigeria.   2015: Still in service.

LANMAR SHORE 389195 IMO 8104278 TM tug/mooring tender; FiFi	118 14.09.1981 29.10.1981	133	22,81 6,61 3,43	Caterpillar 1450bhp 2 x 12-cyl 10.0 knots	Land and Marine Engineering Ltd, Bromborough

08.1981: Registered at Liverpool.   1993: Sold to Costain Oil, Gas & Process Ltd, Wythenshaw, Greater Manchester.   2006: Sold to Forcados Lineboats Ltd for use at Forcados Oil Terminal, Niger Delta, Nigeria.   2015: Still in service.

STIRLING TEAL 701012 IMO 8110617 TM offshore supply ship	119 13.01.1982 28.04.1982	840 351 1175	57,21 11,60 4,65	Yanmar Diesels 3200bhp 2 x 6-cyl 12.5 knots	Harrisons (Clyde) Ltd, Glasgow

05.1982: Registered at Glasgow.   1989: Sold to Stirling Offshore Ltd, Glasgow.   11.1994: Sold to Lamnalco (Cyprus) Ltd, Sharjah, United Arab Emirates (UAE).   Glasgow registry closed.   Registered at Limassol, Cyprus, as **LAMNALCO TEAL**.   10.2001: Sold to Land and Marine National Contracting Co (W.L.L.), Sharjah, UAE.   11.2002: Sold to Warner Offshore Corp, c/o Seaport International Shipping Co LLC, Sharjah, UAE.   11.2002: Limassol registry closed.   Registered at Panama.   2003: Registered at Panama as **PACIFIC**.   12.2007: Sold to Humaid Badir Marine Shipping, c/o Seaport International Shipping Co LLC, Sharjah, UAE.   2015: Still in service.

STIRLING SNIPE 701023 IMO 8110629 TM offshore supply ship	120 27.04.1982 19.07.1982	840 351 1175	57,21 11,60 4,65	Yanmar Diesels 3200bhp 2 x 6-cyl 12.5 knots	Stirling Shipping Co Ltd, Glasgow

07.1982: Registered at Glasgow.   1994: Sold to Lamnalco (Cyprus) Ltd, Sharjah, United Arab Emirates (UAE).   Glasgow registry closed. Registered at Limassol, Cyprus as **LAMNALCO SNIPE**.   01.2006: Sold to Marine Offshore Services Ltd, c/o of Whitesea Shipping & Supply LLC, Sharjah, UAE.   03.2006: Limassol registry closed.   Registered Moroni, Comoros, as **SHAHZADEH**.   12.03.2007: Sold to Gide Marine Services Ltd, c/o Whitesea Shipping and Supply (LLC), Sharjah, UAE.   03.2007: Moroni registry closed.   Registered at Kingstown, St Vincent & the Grenadines.   2015: Still in service.

NORBRIT FAITH 389096 IMO 8116623 Motor cargo ship	121 20.08.1982 07.12.1982	1597 1015 2387	65,03 13,01 6,02	A P E Allen 1350bhp 8-cyl 11.0 knots	North British Shipping Ltd, Hull

10.12.1982: Registered at Hull.   31.05.1983: Alteration to engine particulars - 999bhp.   29.03.1985: Sold to St Christopher Steam Fishing Co Ltd, Hull.   11.1985: Sold to Norbrit Maas B.V, Rotterdam, Netherlands.   28.11.1985: Hull registry closed.   Registered at Rotterdam as **NORBRIT MAAS**.   1987: Sold to Santa Monica Ltd, Nassau, Bahamas.   Rotterdam registry closed.   Registered at Nassau as **NATACHA**. 1989: Sold to Transportes Maritimos Internacionais Ltda, Lisbon, Portugal.   1990: Registered at Nassau as **NATACHA CALDAS**. 01.1991: Sold to Carisbrooke Shipping PLC, Cowes, Isle of Wight.   Nassau registry closed.   Registered at Cowes as **NATACHA C**. 1994: Cowes registry closed.   Registered at Bridgetown, Barbados.   Remeasured to 1636gt, 871net, 2367dwt.   1996: Additiional engine fitted: 6-cylinder 998bhp oil engine by Krupp Mak Maschinenbau G.m.b.H, Kiel - 10 knots.   05.1997: Sold to Seatech A/S, Bergen. Bridgetown registry closed.   Registered at Nassau.   06.1997: Sold to Misje Offshore Marine A/S, Bergen with benefit of 12 months back charter to Carisbrooke.   06.1998: Sold to A/S AMI, Bergen.   Registered at Nassau as **NATACHA**.   06.2004: Sold to Argonafplia Maritime Co, Elefsis, Greece.   Nassau registry closed.   Registered at Eleusis as **VASSILIKI T**.   04.08.2009: In the Kafireas Strait, Andros Island, in collision with bulk carrier JIN ZHOU (27993gt/2001).   Both vessels arrived off Karystos for survey.   Suffered damage to shell plating. 01.2013: Sold to Adriatic Shipping Co Ltd, c/o Charalampos Maritime, Eleusis (Mediteranean Navigation Pvt Ltd, Mumbai, India, managers). Piræus registry closed.   Registered at Belize as **OLYMPIC LIGHT**.   2015: Still in service.

Lady Constance (115) at Goole and awaiting formal handover.

(Bernard McCall)

Lanmar Shore (118) on trials in the Humber.

(Cochrane archive)

Stirling Teal (119)

(Cochrane archive)

NORBRIT HOPE	122	1597	65,03	A P E Allen	North British Shipping Ltd,
701515 IMO 8116635	17.11.1982	1014	13,01	1369hp 8-cyl	Hull
Motor cargo ship	10.03.1983	2380	6,02	11.0 knots	

17.03.1983: Registered at Hull.   29.03.1985: Sold to Lowestoft Fish Selling Co Ltd, Lowestoft.   11.1985: Sold to Norbrit Rijn B.V, Rotterdam, Netherlands.   28.11.1985: Hull registry closed.   Registered at Rotterdam as **NORBRIT RIJN**.   1987: Sold to Santa Monica Ltd, Nassau, Bahamas.   Rotterdam registry closed.   Registered at Nassau as **CATARINA**.   1989: Sold to Transportes Maritimos Internacionais Ltda, Lisbon, Portugal.   1990: Registered at Nassau as **CATARINA CALDAS**.   02.1991: Sold to Carisbrooke Shipping PLC, Cowes, Isle of Wight.   Nassau registry closed.   Registered at Cowes as **CHERYL C**.   1994: Cowes registry closed.   Registered at Bridgetown, Barbados.   Remeasured to 1636gt, 871net, 2367dwt.   08.04.2001: Whilst on passage from Porto Nogaro, Italy, towards Goole with a cargo of steel, stranded on rocks near Cabo Carvoeiro on the Peniche peninsula, approx 60 miles north of Lisbon, due to a navigational error.   Came afloat and subsequently sank in shallow water; all crew rescued.   2001: Registry closed.

STRIDENCE	123	698	81,01	Aabenraa	London & Rochester Trading
702709 IMO 8203244	30.03.1983	439	11,41	1000bhp 8-cyl	Co Ltd (Crescent Shipping),
Motor cargo ship	24.05.1983	1842	5,41	10.0 knots	Rochester

05.1983: Registered at London.   1987: Sold to Crescent Shipping Ltd, Rochester.   1993: London registry closed.   Registered at Nassau, Bahamas.   01.07.1994: Remeasured to 1426gt, 497net, 2088dwt.   11.1997: Sold to Crescent Navigation Ltd, Fareham.   2000: Sold to Crescent Tankships Ltd, Southampton.   01.2002: Sold en bloc to clients of Arpa Shipping BV, Roosendaal, with TURBULENCE (see yard no.124), URGENCE (see yard no.112) and VIBRENCE (see yard no.113).   Later sold to m.v. Muriel Ltd, Kherson, Ukraine.   Registered at Nassau as **MURIEL**   2006: Sold to Muriel Ltd, Roosendaal, Netherlands.   06.2008: Sold to Arpa Shipping BV, Roosendaal.   10.2008: Sold to Tramontane Shipping Co Ltd, Klundert.   Registered at Nassau as **TRAMONTANE**.   09.2010: Sold.   Registered at Nassau as **TAMARA C**. 08.2011: Sold to Defne S Shipping Ltd c/o Danmar Shipmanagement NV, Istanbul.   01.09.2011: Nassau registry closed.   Registered at Panama as **DEFNE K**.   30.12.2013: Broken up.   Registry closed.

TURBULENCE	124	699	81,01	Aabenraa	London & Rochester Trading
702717 IMO 8203256	30.03.1983	440	11,41	1000bhp 8-cyl	Co Ltd (Crescent Shipping),
Motor cargo ship	24.05.1983	1842	5,41	CPP 10.0 knots	Rochester

24.08.1983: Registered at London.   1987: Sold to Crescent Shipping Ltd, Rochester.   09.1992: London registry closed. 01.10.1992: Registered at Nassau, Bahamas.   01.07.1994: Remeasured to 1426gt, 497net, 1821dwt.   11.1997: Sold to Crescent Navigation Ltd, Fareham.   2000: Sold to Crescent Tankships Ltd, Southampton.   2000: Sold to Clipper Wonsild Tankers UK Ltd, Southampton.   01.2002: Sold en bloc to clients of Arpa Shipping BV, Roosendaal, with STRIDENCE (see yard no.123), URGENCE (see yard no.112) and VIBRENCE (see yard no.113).   01.2002: Sold to m.v. Aristote Ltd, Kherson, Ukraine, c/o SMS Ship Management Support BV, Klundert, Netherlands.   01.2002: Registered at Nassau as **ARISTOTE**.   2015: Still in service.

WILLONIA	125	799	73,03	Krupp MaK	J Wharton (Shipping) Ltd,
364589 IMO 8213407	25.10.1983	585	12,60	1285bhp 6-cyl	Gunness
Motor cargo ship	18.01.1984	2415	6,30	CPP 10.0 knots	

23.02.1984: Registered at Goole.   26.06.1986: Sold to F T Everard Shipping Ltd, Greenhithe.   06.05.1988: Registered at Goole as **SANGUITY**.   29.10.1992: On passage Aalborg towards Larne, in collision with Hals Barre Light and shell plate holed. Effected temporary repairs at Hals harbour.   04.11.1992: Sailed for Larne.   05. 2001: Sold to Charles M Willie & Co (Shipping) Ltd, Cardiff.   Goole registry closed.   Registered at Cardiff as **CELTIC FORESTER**.   Engine re-rated to 999bhp.   04.2006: Sold to THH Shipinvest VI ApS, Svendborg, Denmark.   05.2006: Sold to THH Shipinvest V ApS, Svendborg.   Cardiff registry closed.   Registered at St John's, Antigua & Barbuda, as **PIA STEVNS**.   2009: Registered at St John's as **OLIVER STEVNS**.   2012: Laid up at Svendborg.   11.07.2012: Broken up by Jatop ApS, Frederikshavn (Orla's Productforretning).   Registry closed.

SELECTIVITY	126	799	73,03	Krupp MaK	Investors in Industry PLC,
705631 IMO 8213419	21.02.1984	585	12,00	1020bhp 6-cyl	London
Motor cargo ship	05.05.1984	2415	6,30	CPP 10.0 knots	

05.1984: Registered at London. Leased to F T Everard & Sons Ltd, Greenhithe.   1989: Company restyled 3i's PLC, London.   06.1999: Sold to Commercial Acquisitions Ltd. Continued to be leased to F T Everard Shipping Ltd.   05.2000: Sold to F T Everard & Sons Management Ltd.   05.2000: Sold to Scottish Navigation Co Ltd, London.   05.2001: Sold to Charles M Willie & Co (Shipping) Ltd, Cardiff. London registry closed.   Registered at Cardiff as **CELTIC PIONEER**.   05. 2006: Sold to THH Shipinvest V ApS Svendborg, Svendborg, Denmark. Cardiff registry closed.   Registered at Saint John's, Antigua & Barbuda, as **NURAL STEVNS**.   2012: Laid up at Svendborg.   30.06.2012: Broken up by Jatop ApS, Frederikshavn (Orla's Productforretning).   Registry closed.

STEVONIA	127	799	73,00	Krupp MaK	J Wharton (Shipping) Ltd,
364590 IMO 8214346	29.11.1985	585	12,60	1285bhp 6-cyl	Gunness
Motor cargo ship	25.02.1986	2415	6,30	CPP 10.0 knots	

06.03.1986: Registered at Goole.   26.06.1986: Sold to F T Everard Shipping Ltd, Greenhithe.   12.02.1987: Registered at Goole as **SOCIALITY**.   1994: Sold to F T Everard & Sons Management Ltd, Goole.   01.07.1994: Remeasured to 1892gt, 2887dw.   12.2000: Sold to Joint Stock Company 'Limarko', Klaipeda, Lithuania.   Goole registry closed.   Registered at Klaipeda as **SIUITA**.   03.2006: Sold to UAB 'Limarko' (JSC 'Limarko'), Klaipeda.   Engine re-rated to 999bhp.   12.2007: Sold to Stevns Trader Ltd, Saint Johns, Antigua & Barbuda, (part of THH Shipping Ltd, Svendborg, Denmark).   Klaipeda registry closed.   Registered at St John's as **NIELS STEVNS**.   13.02.2012: Sold to Kilic Deniz, Milas, Turkey.   St John's registry closed.   01.03.2012: Registered at Istanbul as **KILIC 1**.   Converted to a live fish carrier. 2015: Still in service.

STAR ALTAIR	128	1704	61,02	W H Allen	Star Offshore Services
364590 IMO 8303032	29.08.1984	654	15,51	5000bhp 2 x 6-cyl	Marine Ltd,
TM supply ship,	02.05.1985	2234		CPP 12.0 knots	Aberdeen
FiFi					

05.1985: Registered at Aberdeen.   1994: Aberdeen registry closed.   Registered at Bridgetown, Barbados.   01.1996: Sold to Stirling Offshore Ltd, Glasgow.   Registered at Bridgetown as **STIRLING ALTAIR**.   12.01.2002: Sold to Boston Putford Offshore Safety Ltd, London. 04.2002: Renamed **PUTFORD ENTERPRISE**.   05.2002: Bridgetown registry closed.   Registered at Lowestoft.   05.2002: Upgraded to survivor multi-role standby safety vessel.   01.07.2002: Remeasured to 1793gt, 2270dwt.   2015: Still in service.

SEAMAN 390365 IMO 8317643 TM tug/ anchor handler	129 26.11.1984 18.02.1985	527 158	32,01 10.61 5,92	Ruston Diesels 4660bhp 2 x 8-cyl 12.5 knots	United Towing (Keith) Ltd, Hull

04.05.1984: Keel laid.   07.1984: Measured.   01.03.1985: Registered at Grimsby.   1986: Owners restyled Humber Tugs (Keith) Ltd, Hull.
24.03.1988: Sold to Humber Tugs Ltd, Immingham.   11.1990: Sold to Rimorchiatori Riuniti Porto di Genova S.r.L, Bari, Italy.
05.12.1990: Grimsby registry closed.   Registered at Bari as **GENUA**.   03.1997: Sold to Classic Marine Ltd, Honduras.   Bari registry closed.
01.04.1997: Registered at Kingstown, Saint Vincent & the Grenadines, as **LADY HAMMOND**.   2000: Remeasured to 498gt, 149net.
18.10.2006: Sold to Marine Capabilities LLC, Abu Dhabi, United Arab Emirates.   10.2006: Kingstown registry closed.   Registered at Abu Dhabi.   2015: Still in service.

HEBRIDEAN ISLES 711814 IMO 8404812 TM RoRo / cargo passenger ship	130 04.07.1985 27.11.1985	3040 912	78,03 15,80	Mirrlees Blackstone 4690bhp 2 x 8-cyl 15.0 knots	Caledonian MacBrayne Ltd, Glasgow

First MacBrayne vessel to be built outside Scotland.
12.1985: Registered at Glasgow.   05.12.1985: Entered service.   06.12.1985: First commercial sailing: Stornoway-Ullapool service.
04.10.2006: Managed by Calmac Ferries Ltd, Gourock.   2006: Company restyled Caledonian Maritime Assets Ltd, Port Glasgow.
2007: Leased to CalMac Ferries Ltd.   2015: Still in service.

Not built	131 Order cancelled.

STIRLING ESK 711822 IMO 8501505 TM supply ship, FiFi	132 15.11.1985 02.04.1986	1399 572	58,00 14,02	Yanmar Marine 5132bhp 2 x 8-cyl 2 x AZP 12.0 knots	Harrisons (Clyde) Ltd, Glasgow

Built at Goole.   04.1986: Registered at Glasgow.   1996: Sold to Stirling Offshore Ltd, Glasgow.   10.2000: Glasgow registry closed.
Registered at Douglas, Isle of Man.   12.2003: Sold to Stirling Offshore Ltd, c/o Boston Putford Offshore Safety Ltd, Lowestoft.   Douglas registry closed.   Registered at Lowestoft as **PUTFORD TERMINATOR**.   2003: Upgraded to survivor multi-role standby safety vessel - 1504gt, 2151dwt.   2015: Still in service.

WIMPEY SEASPRITE 709908 IMO 8503515 TM supply ship	133 09.05.1986 15.07.1986	1001 385 1442	54,05 12,80 5,20	MAN-B&W Diesel 3590bhp 2 x 6-cyl CPP 12.0 knots	Wimpey Marine Ltd, Great Yarmouth

07.1986: Registered at London.   07.1988: Sold to Offshore Logistics Marine Ltd, London.   Registered at London as **HIGHLAND SPRITE**.
1989: Sold to Gulf Offshore Marine International Inc, Panama.   11.1990: Sold to Gulf Offshore N.S. Ltd, Aberdeen & London.
11.2002: London registry closed.   Registered at Panama.   20.03.2009: Sold to Seamar Shipping BV, Den Helder, Netherlands.   Panama registry closed.   25.03.2009: Registered at Gibraltar as **SEAMAR SPLENDID**.   Converted to a research/diving support vessel - 1199gt, 1297dwt.   2015: Still in service.

WIMPEY SEAWITCH 709907 IMO 8503527 TM supply ship	134 25.04.1986 21.08.1986	1001 385 1442	54,05 12,80 5,20	MAN-B&W Diesel 3590bhp 2 x 6-cyl CPP 13.25 knots	Wimpey Marine Ltd, Great Yarmouth

Built by Cochrane, Goole.
08.1986: Registered at London.   1988: Sold to Offshore Logistics Marine Ltd, London.   Registered at London as **HIGHLAND LEGEND**.
11.1990: Sold to Gulf Offshore N.S. Ltd, Aberdeen & London.   11.2002: London registry closed.   Registered at Panama.   08.2004: Sold to Gulfmark Asia Pte Ltd, Singapore.   10.2014: Sold to Amiran Mje, c/o Baltic Marine Services, Dubai, United Arab Emirates.   Renamed **MATAF DENALI**   2015: Still in service.

ST. CECILIA 712850 IMO 8518546 M RoRo cargo/ passenger ferry	135 04.11.1986 24.03.1987	2968 904	77,05 55,20	H&W-MAN 3267bhp 3 x 6-cyl 3 x VSP 12.0 knots	Eurolease Corp Ltd, London

Chartered to Sealink Isle of Wight Ferries - Portsmouth-Fishbourne service.   03.1987: Registered at London.   1990: Transferred to Wightlink Ltd, Portsmouth (Isle of Wight Ferries), London.   1995: Sold to Wightlink Ltd, Portsmouth.   2015: Still in service.

LADY THERESA 712666 IMO 8616245 TM tug	136 25.09.1987 19.11.1987	156	22,10 7,32	Ruston Diesels 1900bhp 2 x 6-cyl 10.5 knots	Humber Tugs (Jellicoe) Ltd, Immingham

26.01.1988: Registered at Hull.   24.03.1988: Sold to Humber Tugs Ltd, Immingham.   1994: Sold to Clyde Shipping Co Ltd, Glasgow (Tyne & Wear Tugs Ltd, Sunderland).   Hull registry closed.   Registered at Newcastle as **HILLSIDER**.   Fore end rebuilt.   05.1995: Company and assets sold to Cory Towage Ltd, Woking.   03.2000: Company and assets sold to Wijsmuller BV, IJmuiden, for the sum of £82 million.   Newcastle registry closed.   Registered at Rotterdam.   09.2000: Sold to Fairplay Schleppdampfschiffs-Reederei Richard Borchard GmbH, Hamburg.   Registered at Rotterdam as **FAIRPLAY XI**.   Working at Rotterdam.   2003: Rotterdam registry closed.   Registered at Szczecin, Poland.   12.2013: Sold to Lührs Schifffahrt GmbH & Co KG, Hamburg.   Szczecin registry closed.   Registered at Hamburg as **TWISTER**.   2015: Still in service.

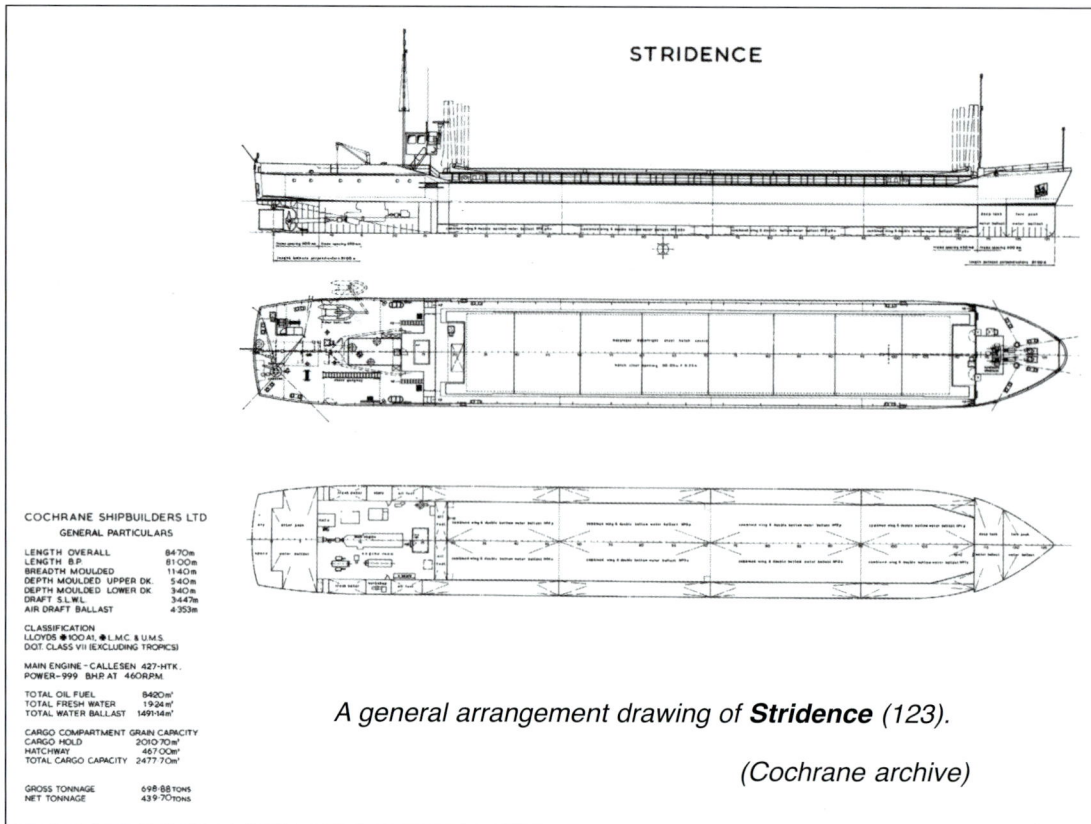

STRIDENCE

COCHRANE SHIPBUILDERS LTD
GENERAL PARTICULARS

LENGTH OVERALL	84·70m
LENGTH B.P.	81·00m
BREADTH MOULDED	11·40m
DEPTH MOULDED UPPER DK.	5·40m
DEPTH MOULDED LOWER DK.	3·40m
DRAFT S.L.W.L.	3·447m
AIR DRAFT BALLAST	4·353m

CLASSIFICATION
LLOYDS ✠ 100 A1, ✠ L.M.C. & U.M.S.
D.O.T. CLASS VII (EXCLUDING TROPICS)

MAIN ENGINE – CALLESEN 427-HTK.
POWER–999 B.H.P. AT 460 R.P.M.

TOTAL OIL FUEL	84·0m³
TOTAL FRESH WATER	19·24m³
TOTAL WATER BALLAST	1491·14m³

CARGO COMPARTMENT GRAIN CAPACITY
CARGO HOLD	2010·70m³
HATCHWAY	467·00m³
TOTAL CARGO CAPACITY	2477·70m³

GROSS TONNAGE	698·88 TONS
NET TONNAGE	439·70 TONS

*A general arrangement drawing of **Stridence** (123).*

(Cochrane archive)

Hebridean Isles (130) *nearing completion.*

(Roy Cressey collection)

Wimpey Seasprite (133) *is also nearing completion.*

(Roy Cressey collection)

Seaman (129) on trials for United Towing.

(Cochrane archive)

Seaman later in her career in the colours of West Coast Towing (UK) Ltd and named **Lady Hammond**. In this photograph, she is clearly registered at Swansea but this was very short-lived.

(Danny Lynch)

LADY JOAN 712669 IMO 8616257 TM tug	137 23.10.1987 21.08.1988	156	22,10 7,32	Ruston Diesels 1926bhp 2 x 6-cyl 10.5 knots	Humber Tugs (Sturdee) Ltd, Immingham	

02.02.1988: Registered at Hull. 24.03.1988: Sold to Humber Tugs Ltd, Immingham. 11.1992: Sold to Rimorchiatori Ancona SpA, Ancona, Italy. 04.11.1992: Hull registry closed. 05.11.1992: Registered at Ravenna as **CONERO**. Classed as a coastal tug; 143grt. 2015: Still in service.

LADY SYBIL 712668 IMO 8616269 TM tug	138 08.10.1987 01.1988	156	22,10 7,32	Ruston Diesels 1900bhp 2 x 6-cyl 10.5 knots	Humber Tugs (Beatty) Ltd, Immingham	

Built at Goole; final Goole-built ship. 26.01.1988: Registered at Hull. 24.03.1988: Sold to Humber Tugs Ltd, Immingham. 1998: Sold to Howard Smith (Humber) Ltd, Hull. 1998: Hull registry closed. 02.1999: Sold to Fairplay Schleppdampfschiffs-Reederei Richard Borchard GmbH, Hamburg. 03.1999: Registered at Hamburg as **FAIRPLAY X**. Operating at Rotterdam. 07.06.2007: Sold to Marine & Towage Services Group Ltd, Falmouth. Hamburg registry closed. Registered at Falmouth as **MTS VENGEANCE**. Engines re-rated to 2100bhp total. 2015: Still in service.

LADY DULCIE 712667 IMO 8616271 TM tug	139 13.06.1988 01.08.1988	157	22,10 7,32	Ruston Diesels 1926bhp 2 x 6-cyl 10.5 knots	Humber Tugs Ltd, Immingham	

Laid down as **LADY THELMA**. 11.08.1988: Registered at Hull as **LADY DULCIE**. 01.1990: Sold to Rimorchiatori Riuniti Panfido E Compagnia S.R.L, Trieste, Italy. 29.01.1990: Hull registry closed. 02.1990: Registered at Trieste as **FULGOR**. 02.1992: Sold to Società Rimorchi e Salvataggi -Trieste a.r.l, Trieste. 1993: Sold to Tripnavi S.p.A, Trieste. 02.1995: Sold to Tripmare S.r.l., Trieste. 02.1997: Sold to Sociedad Naviera Ultragas Ltd, Italy. 04.1997: Sold to Petrolera Transoceanica S.A, Lima, Peru. Trieste registry closed. 06.1997: Registered at Lima as **SAN GALLAN**. Based at Lima. 01.2015: Sold to Abadia Delmar SA. 2015: Still in service.

MARIA ISABEL I 14058-L1 IMO 8618334 Motor tug/supply, anchor handler, FiFi	140 28.05.1987 24.07.1987	376	28,02 9,63	Ruston Diesels 4340bhp 2 x 6-cyl 2 x AZP 14.0 knots	Cory Towage (Bermuda) Ltd, Panama	

21.01.1987: Keel laid as **MARIA ISABEL**. 07.1987: Registered at Panama as **MARIA ISABEL I**. 06.1997: Sold to Cory Towage Ltd, Woking. 05.03.1998: Sold to Stevns Towage & Charter and others, c/o Nordane Shipping A/S, Svendborg, Denmark. Panama registry closed. Registered at Svendborg as **STEVNS BUGSER**. 30.11.2000: Sold to A/S Em Z Svitzer, Copenhagen. 01.2001: Svendborg registry closed. 23.01.2001: Registered at Kalundborg as **EGIL**. 2012: Sold to Svitzer Canada Ltd, Halifax, Nova Scotia, Canada. 01.06.2012: Svendborg registry closed. Registered at Panama. 2015: Still in service.

MARIA LUISA II 14094-L1 IMO 8618334 Motor tug/supply, anchor handler, FiFi	141 12.08.1987 02.10.1987	376	28,02 9,63	Ruston Diesels 4340bhp 2 x 6-cyl 2 x AZP 14.0 knots	Odyssey Towage (Bermuda) Ltd, Panama	

28.01.1987: Keel laid as **MARIA LUISA**. 10.1987: Registered at Panama as **MARIA LUISA II**. 08.1998: Sold to Cory Towage Ltd, Woking. Panama registry closed. Registered at Liverpool. 01.2000: Sold to to Svitzer Marine UK Ltd. Liverpool registry closed. Registered at Kingstown, St Vincent & the Grenadines. 2002: Chartered to Kompania di Korsou, Willemstad. Kingstown registry closed. Registered at Willemstad. 04.2006: Sold to Plipwijs Ltd, Port of Spain, Trinidad & Tobago. Willemstad registry closed. Registered at Port of Spain. 2008: Sold to Svitzer (Americas) Ltd, Trinidad. 08.2008: Port of Spain registry closed. Registered at Kingstown, St Vincent & the Grenadines. 03.2010: Charter completed. 03.2010: Sold to Svitzer Andino S.A., Lima, Peru. Kingstown registry closed. Registered at Lima. 2011: For sale. 12.2014: Sold to Abadia Del Mar SA, Buenos Aires, Argentina. Lima registry closed. 01.2015: Registered at Buenos Aires. 2015: Still in service.

ALLEGIANCE S 709224 RSS A22669 SH90 Motor trawler	142 31.03.1987 05.1987	120 31	16,32 6,71 4,02	Kelvin Diesels 440bhp 6-cyl 8.75 knots	T & J Sheader & F & D Normandale, Scarborough	

Built by Cochrane, Goole. Total cost £510,000.
Ordered by Frederick George Normandale & Thomas Sheader, Scarborough, from Steelships Ltd, Truro, for £460,000. 1986: Watercraft Group, the parent company of the builder, in receivership. Material already worked cut up and transported by Cochrane, with other ordered material, to Goole for re-assembly and build. 22.05.1987: Entered service. Registered at Scarborough (SH90). 21.12.2006: Sold to SH90 Ltd, Scarborough (Frederick George Normandale & Danny Normandale, Scarborough). 01.12.1988: RSS No.A22669 allotted.
2011: Registered at Scarborough as **ALLEGIANCE** (SH90). 10.2011: Arrived Parkol Marine Engineering Ltd, Whitby, for lengthening by 7,44m and modernisation at a cost of approximately .£300,000. 15.12.2011: New Record of Particulars issued following final survey - 23,76m x 6,72m x 4,02 m; 206gt. Engine 309kW/414hp. 2015: Still in service fishing out of Peterhead.

THORNELLA 709224 IMO 8705216 H96 Motor stern trawler	143 21.12.1987 10.03.1988	555 166	33,50 9,72 4,25	Ruston Diesels 1724bhp 6-cyl 12.0 knots	J Marr Ltd, Hull	

Contract price £2.5million.
08.1987: Keel laid. 10.1987: Measured. 09.03.1988: Registered at Hull (H96). 01.12.1988: RSS No.A17736 allotted. 06.1996: Sold to Thornella Ltd, Hull. 03.10.2003: Sailed from Hull on final trip. 24.10.2003: Landed in Germany. 11.2003: Sold en bloc with LANCELLA (see yard no.144) to Baines Bros Ltd, New Holland, for breaking up under English Decommissioning Scheme. 27.11.2003: Delivered to New Holland. 12.2003: Hull registry closed.

A panoramic view of the launch of the **Maria Isabel** (140).

(Roy Cressey collection)

Maria Luisa II (141) at Newport.

(Danny Lynch)

The buff hull of the **Thornella** (143) stands out against the gloomy background.

(Roy Cressey collection)

139

LANCELLA 712683　IMO 8705228 RSS A17738 H98　　Motor stern trawler	144 18.04.1988 06.1988	555 166	33,50 9,72 4,25	Ruston Diesels 1724bhp　6-cyl 12.0 knots	J Marr Ltd, Hull

Contract price £2.5 million.

11.1987: Keel laid.　04.1988: Measured.　24.06.1988: Registered at Hull (H98).　01.12.1988: RSS No. A17738 allotted.　04.1998: Sold to Lancella Ltd, Hull.　26.01.1990: With Sk Dave Wright achieved a British record grossing of £260,324 for a single trip.　02.04.2003: Sailed from Scrabster on final trip.　12.04.2003: Landed at Hull and laid up.　10.2003: Sold en bloc with THORNELLA (see yard no.143) to Baines Bros Ltd, New Holland, for breaking up under English Decommissioning Scheme.　28.10.2003: Delivered to New Holland.　12.2003: Hull registry closed.

Un-named Flat-topped open barges	145, 146, 147 12.1986		24,38 9,02		

Launched at Goole.

GLENROSE I 811108　IMO 8821010 H2　　Motor stern trawler	148 25.06.1990 03.10.1990	554 166	33,50 9,72 4,25	Ruston Diesels 1724bhp　6-cyl 12.0 knots	Onward Fishing Co Ltd, Hull

09.1990: Registered at Hull (H2).　RSS No.B11108.　09.1999: Sold to J Marr (Shipping) Ltd, Hull.　1999: Registered at Hull as **KIRKELLA** (H2).　29.09.1999: Sailed on first trip for new owners.　03.2001: Sold to Kirkella Ltd, Hull.　12.2006: Sold to Mackinco Ltd, Torshavn, Faroe Islands.　2006: Registered at Hull as **GLEN ROSE II**.　2007: Sold to Black Island Shipping Co Ltd, Aberdeen.　02.2009: Sold to Bradava,Ventspils, Latvia, and converted to a pelagic trawler.　2009: Hull registry closed.　Registered at Ventspils as **GLENROSE** (LVV1555).　2015: Still in service.

Not built	149　Order cancelled				

POWERFUL 710490　IMO 8705400 　　TM tug, FiFi, 　　pollution control	150 07.03.1988 04.05.1988	308 94 258	27,12 9,33	Ruston Diesels 3200bhp　2 x 6-cyl 12.0 knots	Bermuda Government Marine & Ports, Hamilton, Pembroke, Bermuda

05.1988: Registered at Hamilton, Bermuda.　14.05.1988: Sailed from shipyard on passage to Bemuda.　05.2014: Completed major refit at Jacksonville, Florida.　2015: Still in service.

FAITHFUL 715329　IMO 8705412 　　TM tug, FiFi, 　　pollution control	151 04.08.1989 07.03.1990	302 92 258	27,04 9,33	Ruston Diesels 3198bhp　2 x 6-cyl 12.0 knots	Bermuda Government Marine & Ports, Hamilton, Pembroke, Bermuda

04.1990: Registered at Hamilton, Bermuda.　02.04.1990: Sailed from shipyard on passage to Bermuda.　09.2012: Major engine overhaul.　2015: Still in service.

Not built	152, 153　Orders cancelled				

BRABOURNE 717041　IMO 8801321 　　Motor oil products 　　tanker	154 30.09.1988 13.12.1988	1646 846 2675	73,09 12,62 6,00	Krupp MaK 1795bhp　6-cyl 10.0 knots	Lloyds Plant Leasing Ltd, London

Leased to Bowker & King Ltd, London.

01.1989: Registered at London.　1992: Sold to Lloyds International Leasing Ltd, London.　11.1998: Sold to Crescent Navigation Ltd, Southampton.　01.10.2002: Remeasured to 1646gt, 2524dwt.　06.2006: Registered at London as **CLIPPER BRABOURNE**. 16.07.2007: Sold to Polrom Oil Maritime Co, Piræus, Greece.　07.2007: Registered at London as **AKTEA OSRV**.　01.08.2007: London registry closed.　Registered at Piræus.　01.08.2008: Converted to a pollution control vessel.　2015: Still in service.

BLACKROCK 717140　IMO 8807595 　　Motor oil products 　　tanker	155 08.02.1989 05.04.1989	1646 846 2675	73,09 12,62 6,00	Krupp MaK 1795bhp　6-cyl 10.0 knots	Lloyds Plant Leasing Ltd, London

Leased to Bowker & King Ltd, London.

04.1989: Registered at London.　1992: Sold to Lloyds International Leasing Ltd, London.　11.1998: Sold to Crescent Navigation Ltd, Southampton.　01.2007: Sold to John H Whitaker (Tankers) Ltd, Hull.　Registered at London as **WHITDAWN**.　02.2011: Sold to ECR Leasing Services SA de CV c/o Bunkers Mexico Energy SAPI de CV, Mexico City, Mexico.　03.2011: London registry closed.　Registered at Mexico City as **GOLA I**.　2015: Still in service, employed as a bunkering tanker.

DANIELLE OF HAMM 　　Charter yacht	156 1990	120 displacement 135gt	32,70 6,58 2,60 draught	Perkins Sabre Twin 160bhp 10.0 knots max	Jankel Schooners

Designed by Laurent Giles Naval Architects Ltd, Lymington, Hampshire.　1998 and 2002: Refitted at Lymington.　2002: Renamed **TIGERLILY OF CORNWALL**.　2008: Refitted and updated.　2015: Still in service.

Not built	157 - 161 inclusive.　Orders cancelled.				

The launch of the **Glenrose I** (148).

(Roy Cressey collection)

The later stages in fitting out of the **Faithful** (151).

(Andrew Wiltshire collection)

With the tug **Keelman** in attendance, the **Danielle of Hamm** (156) is lifted into the River Ouse by crane. She was the only vessel to be launched in this way at the Selby yard.

(Roy Cressey collection)

VARAGEN	162	928	50,00	Caterpillar Inc,	British Linen Leasing Ltd,
710153 IMO 8818154	07.04.1989	312	11,70	2148bhp 2 x 12-cyl	Edinburgh
TM RoRo vehicle	15.06.1989	321	4,50	15.0 knots	
passenger ferry					

06.1989: Registered at Kirkwall.   Demise chartered to Orkney Ferries Ltd, Kirkwall (Orkney Islands Council Ferry Services, Kirkwall).
08.2003: Sold to Capital Leasing Ltd, Edinburgh. Charter continued.   08.2012: Sold to Orkney Islands Council, Kirkwall.   2015: Still in service.

UNION JUPITER	163	2230	96,00	Alpha MAN B&W	Union Transport Group plc,
717498 IMO 8901030	16.11.1989	1244	12,50	1018bhp 8-cyl	Bromley
Motor cargo ship	21.03.1990	3274	6,35	CPP 11.2 knots	

05.1990: Registered at London.   1996: London registry closed.   Registered in Bridgetown, Barbados.   2006: Bridgetown registry closed.
Registered at Douglas, Isle of Man.   05.2009: Sold to Kalhim Shipping Inc, Istanbul, Turkey, c/o MCE Kargo-Mahmut Can Egerci, Istanbul,
Turkey.   Registered at Bridgetown as **BLUES**.   06.2014: Bridgetown registry closed.   06.2014: Registered at Avatiu, Cook Islands.
2015: Still in service.

BROMLEY PEARL	164	2230	96,00	Alpha MAN B&W	Bromley Shipping plc,
713501 IMO 8903026	29.03.1989	1001	12,50	1018bhp 8-cyl	Bromley
Motor cargo ship	18.07.1990	3222	6,35	CPP 11.2 knots	

Hull built by R Dunston (Hessle) Ltd. (yard no.H991).
08.1990: Registered at Harwich.   07.1995: Sold to Union Transport Group plc, Bromley, Kent.   Harwich registry closed.
13.07.1995: Registered at Bridgetown, Barbados, as **UNION PEARL**.   05.2009: Sold to MCE Kargo-Mahmut Can Egerci, Istanbul, Turkey.
Registered at Bridgetown as **JAZZ**.   21.02.2011: Sold to Makmoll Shipping Inc c/o MCE Kargo-Mahmut Can Egerci, Istanbul.
08.2014: Bridgetown registry closed.   08.2014: Registered at Avatiu, Cook Islands.   2015: Still in service.

UNION SATURN	165	2230	96,00	Alpha MAN B&W	Union Transport Group plc,
N/A IMO 8903038	01.02.1990	1001	12,50	1018bhp 8-cyl	Bromley
Motor cargo ship	1990	3274	6,35	CPP 11.2 knots	

Due for delivery in October 1990 but not ready; contract cancelled by Union Transport.
1990: Laid up in Hull incomplete.   07.1991: Sold to Hardinxveld Boat Charter BV, Gorinchem, Netherlands.   09.1991: Completed by
Richard Dunston (Hessle) Ltd, Hull, (yard no.H992) as **UNION SATURN**.   08.1991: Sold to Tima Shipping & Trading BV, Rotterdam.
30.09.1991: Registered at Limassol, Cyprus, as **TIMA SATURN**.   12.1991: Demise chartered to Carisbrooke Shipping Ltd, Cowes, with
purchase option.   07.1992: Registered at Limassol as **ANJA C**.   02.1994: Sold to MARTESE (Marine Technical Services) Ltd, Limassol
registry closed.   Registered at Kingstown, St Vincent & the Grenadines.   09.1996: Sold to Capital Bank Leasing (9) Ltd, Chester (Medina
Shipping Ltd).   Kingstown registry closed.   Registered at Bridgetown, Barbados.   13.02.2001: Detained by MCA at Fowey with minor defects
and a gauge on one of the cylinders of the fixed firefighting system indicated low. On check all found to be full. MCA declined to remove the
vessel's detention notice from the files.   08.2005: Cowes registry closed.   08.2005: Registered at Valletta, Malta, as **NOVA**.
02.09.2005: Registered to Astrey Shipping Ltd, c/o Sio Shipping Ltd, Baku, Azerbaijan.   2015: Still in service.

UNION MERCURY	166	2230	96,00	Alpha MAN B&W	Union Transport Group plc,
720307 IMO 8903040	05.09.1990	1001	12,50	1018bhp 8-cyl	Bromley
Motor cargo ship	12.12.1990	3222	6,35	CPP 11.2 knots	

30.06.1986: Keel laid.   30.11.1990: Measured.   Subsequently refused by Union Transport due to late delivery.
12.12.1990: Completed as **UNION MERCURY**.   Sold to Dilmun Navigation Co Ltd, London.   12.1990: Registered at Hull.   1991: Demise
chartered to Carisbrooke Shipping Ltd, Cowes, Isle of Wight with option to purchase.   25.06.1991: Registered at Hull as **LESLEY JANE C**.
Engine re-rated 1384bhp.   07.1993: Sold on completion of charter to Carisbrooke Shipping PLC, Cowes.   14.07.1993: Sold to Vectis
Shipping Ltd, Cowes.   26.07.1993: Registered at Hull as **VECTIS ISLE**.   11.1994: Hull registry closed.   Registered at Bridgetown,
Barbados.   11.1997: Sold to Carisbrooke Shipping CV, Rotterdam.   Bridgetown registry closed.   Registered at Rotterdam.
08.2000: Rotterdam registry closed.   Registered at Bridgetown.   Remeasured to 2237gt, 1244net, 3222dwt.   2005: Sold to Vectis Shipping
Ltd, Cowes.   10.2005: Sold to Tonga Management Inc, c/o South River Shipping Co Ltd, Kherson, Ukraine.   Bridgetown registry closed.
Registered at Moroni, Comoros.   2015: Still in service.

SUPERIORITY	167	2230	96,00	Alpha MAN B&W	Scottish Navigation Co Ltd,
719069 IMO 8903052	05.12.1990	1001	12,50	1018bhp 8-cyl	London
Motor cargo ship	08.05.1991	3212	6,35	CPP 11.2 knots	

05.1991: Registered at London.   2002: Sold to Union Transport Group plc, Bromley.   London registry closed.   Registered at Douglas, Isle of
Man, as **UNION EMERALD**.   11.2004: Sold to Silver Maritime Ltd, c/o Pearl Shipping, Baku, Azerbaijan.   Douglas registry closed.
Registered at Valletta, Malta, as **EMERALD**.   10.2007: Sold to Sio Shipping Ltd, Azerbaijan.   09.2008: Sold to Medatlantic Shipping Co Ltd,
Malta.   10.2008: Valletta registry closed.   Registered at Panama as **EMERALDI**.   01.2014: Sold to GN Global Shipping Lines Inc, Istanbul,
Turkey.   Registered at Panama as **LADY ARZU**.   2015: Still in serrvice.

SHORT SEA TRADER	168	2230	96,00	Alpha MAN B&W	Short Sea Europe plc,
719117 IMO 8903064	16.04.1991	1001	12,50	1018bhp 8-cyl	London
Motor cargo ship	26.06.1991	3222	6,35	CPP 11.2 knots	

06.1991: Registered at London.   07.2001: Sold to Union Transport Group plc, Bromley, Kent.   08.2001: London registry closed.
05.08.2001: Registered at Douglas, Isle of Man, as **UNION SATURN**.   08.2006: Sold to Favorite Shipping AS, Bergen, Norway. Back
chartered to Union Transport Group plc, Bromley.   17.07.2013: Sold to Saturn Shipping Ltd, c/o Kargo-Mahmut Can Egerci, Istanbul, Turkey.
07.2013: Douglas registry closed.   Registered at Avatiu, Cook Islands, as **SATURN**.   2015: Still in service.

The **Varagen** (162) in the final stages of fitting out.

(Roy Cressey collection)

The **Union Jupiter** (163) is eased towards the quayside a few minutes after her launch

(Chris Cheetham, Richard Potter collection)

The **Short Sea Trader** (168) on the slipway.

(Roy Cressey collection)

ST. FAITH 718794 IMO 8907228 M RoRo cargo / passenger ferry	169 28.02.1990 20.06.1990	3009 914 594	72,40 24,40 4,50	H&W-MAN 3270bhp 3 x 6-cyl 3 x VSP 12.5 knots	NWS 5 Ltd, London

Chartered to Sealink Isle of Wight Ferries (Sea Containers Ltd, Bermuda) - Portsmouth-Fishbourne service.   07.1990: Registered at London.   11.1991: Transferred by Sea Containers to Wightlink Ltd, Portsmouth (Isle of Wight Ferries), London.   1994: Company and assets sold to management financed by CinVen Ltd, London.   2005: Company and assets sold to Macquarie European Infrastructure Fund, London.   20.03.2010: Stern ramp and fendering damaged when landing heavily on linkspan at Camber Docks, Portsmouth; two passengers suffered minor injuries.   2015: Still in service.

NORTH SEA TRADER 721895 IMO 9011961 Motor cargo ship	170 16.07.1991 25.09.1991	2230 1001 3774	96,00 12,50 6,35	Alpha MAN B&W 1018bhp 8-cyl CPP 11.2 knots	Short Sea Europe plc, London

09.1991: Registered at London.   07.2001: Sold to Union Transport Group plc, Bromley, Kent.   08.2001: London registry closed. 05.08.2001: Registered at Douglas, Isle of Man, as **UNION GEM**.   17.07.2002: Inwards for Ipswich in River Orwell suffered stearing gear failure and ran outside channel into moored yachts and pleasure craft.   08.2006: Sold to Favorite Shipping AS, Bergen, Norway. Back chartered to Union Transport Group plc, Bromley.   17.07.2013: Sold to Neptune Shipping Ltd, c/o Vestra Ltd, Kherson, Ukraine. 07.2013: Douglas registry closed. Registered at Avatiu, Cook Islands, as **NEPTUNE**.   2015: Still in service.

ELIZA P G IMO 9013438 Motor tanker	171 26.11.1991 19.03.1992	3338 1587 5800	90,00 16,00 9,00	Krupp MaK 2991bhp 6-cyl CPP 12.0 knots	Eliza PG Ltd, Shoreham

03.1992: Registered at Douglas, Isle of Man.   05.1995: Sold to Partrederiet Eliza PG DA., Oslo.   2004: Sold to Eliza II PG DA, Oslo. 08.2008: Remeasured to 5440dwt.   09.03.2012: Sold to Bunker Vessel Management SA, Panama City. Douglas registry closed. 03.2012: Registered at Panama as **KELLY TRADER**.   2015: Still in service.

Not built	172 Order cancelled

FORTH BRIDGE IMO 9041136 Motor tanker	173 03.07.1992 10.1992	3338 1744 5800	90,00 16,00 9,00	Krupp MaK 2991bhp 6-cyl CPP 12.0 knots	Forth Tankers plc, Edinburgh

10.1992: Registered at Leith. Chartered by Pritchard-Gordon Tankers Ltd, Shoreham.   08.2008: Sold to Faro Shipping Holdings Ltd, Singapore, for the sum of $7 million.   28.10.2010: Sold to Matrix Shipping Ltd, c/o Val Oil Trading SA, Athens.   10.2010: Leith registry closed.   10.2010: Registered at Monrovia, Liberia.   05.2012: Registered at Lagos as **MATRIX I**.   2015: Still in service.

Following the closure of the yard all equipment, tools, computers, models, paintings etc auctioned on 24, 25 and 26 March 1993.

*The **St Faith** (169) fitting out.*

(Roy Cressey)

LAUNCH PREPARATIONS

Rudderman (1519).

(Roy Cressey collection)

Star Altair (128)

(Cochrane archive)

Hebridean Isles (130)

(Cochrane archive)

LAUNCHES

Kaldbakur *(1322)*

(Cochrane archive)

Kipling *(1417)*

(Cochrane archive)

Kashmir *(1422)*

(Cochrane archive)

Cape Kennedy (1490)

(Cochrane archive)

Cape Kennedy from a different vantage point.

(Cochrane archive)

Pronto (1508)

(Cochrane archive)

Selbydyke (106)

(Cochrane archive)

Stridence (123)

(Cochrane archive)

Eliza PG (171)

(Cochrane archive)

ALEXANDRA TUGS

Waterloo (1389) approaching the lock at Swansea on 8 June 1968.

(John Wiltshire)

Waterloo (1389) and sister tug *Wallasey* (1390) leaving the lock at Swansea on 13 June 1968.

(John Wiltshire)

Canada (1368) and *Formby* (1369) seen at Swansea in December 1969 immediately after being handed over to Italian owners and renamed *Strepitoso* and *Poderoso* respectively.

(Des Harris, Andrew Wiltshire collection)

Danube VII (1312).

(Stuart Emery collection)

Owen Smith (1316).

(Andrew Wiltshire collection)

Hibernia (1483).

(Andrew Wiltshire collection)

Rana (1373).

(C C Beazley, Andrew Wiltshire collection)

IN SERVICE

The **Somersetbrook** (1536) passes Meredyke as she leaves the River Trent. Her external appearance suggests that drydocking could be imminent.

(Bernard McCall collection)

The **Urgence** (112) loads a cargo of grain at Southampton on 13 July 1997

(Krispen Atkinson)

Sistership **Vibrence** (113) was also handling a grain cargo when photographed at Goole.

(John Mattison)

The **Yesso** (1433) had an unusual career once her trawling days were over. We see her near Tower Bridge on 28 April 1984, three years after her conversion to a pollution control vessel named **Cleanseas 1**.

(Roy Cressey)

The **Boston Sea Knight** (1571) arrives at Great Yarmouth on 31 March 1978.

(Bernard McCall)

The **Exegarth** (1568) at Milford Haven on 25 March 1993.

(Andrew Wiltshire)

Now in Wightlink livery, **St Cecilia** (135) is seen at her berth at The Camber in Portsmouth harbour. Located near the entrance, this is the oldest part of Portsmouth's commercial harbour.

(Iain McCall)

Also in Wightlink livery, **St Faith** (169) is about to leave Portsmouth harbour at the start of a voyage to Fishbourne.

(Iain McCall)

In the latter days of the shipyard, a number of prefabricated units were built for Drax power station which is located on the River Ouse between Selby and Goole. The units could be loaded on to the ships that were moored at the building berth only during the rising flood tide. A large hired crane had to be used. Appropriately the ships used had been built at the Selby yard. One of them was the **Selectivity** (126).

(John Adamson)

The other vessel to be used for this work was the **Sociality** (127). Between lifts, the ships had to move to an upstream berth in order to await the next tide at a safe berth, and also before finally sailing to Drax on a young flood tide.

(John Adamson)

MISCELLANY

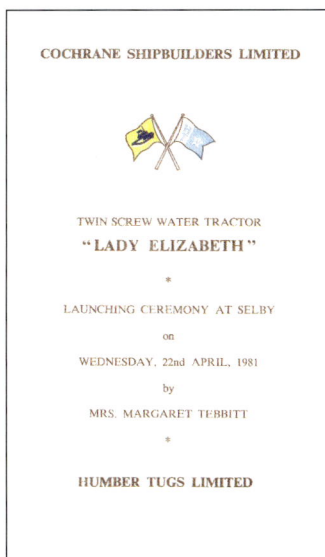

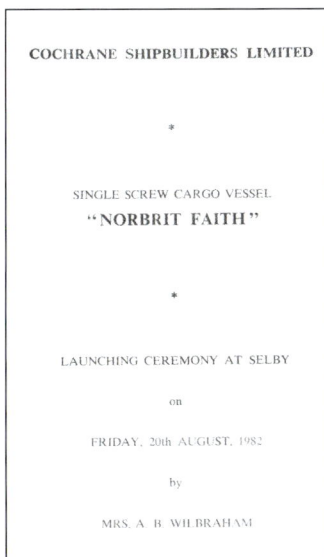

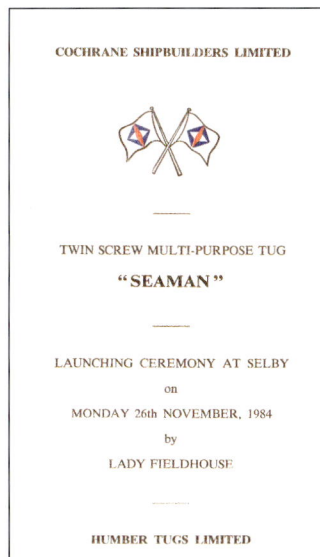

Launch brochures for **Lady Elizabeth** *(114)*, **Norbrit Faith** *(121)* and **Seaman** *(129)*.

(Cochrane archive)

A splendid image of the **Empire Ace** *(1255) as she sails past Old Kilpatrick on passage down the River Clyde.*

(Authors' collection)

The crew of steam trawlers would often have to cope as best they could with coal of very poor quality as evidenced by this photograph of bunkers being loaded on to **Lord Ancaster** *(1337).*

(Authors' collection)

Our final photograph really must be one of a vessel under construction. We see the frames of **Rodney** *(1415).*

(Cochrane archive)

Index of vessel names when launched (and yard numbers)

Index of vessel names following sale or transfer (and yard numbers)